The Organization of Voluntary Movement

Neurophysiological Mechanisms

The Organization of Voluntary Movement

Neurophysiological Mechanisms

Ya. M. Kots
Central Institute of Physical Culture
Moscow, USSR

Translation Edited and with a Foreword by
Edward V. Evarts, M.D.
Chief, Laboratory of Neurophysiology
National Institute of Mental Health
Bethesda, Maryland

SPRINGER SCIENCE+BUSINESS MEDIA, LLC

Library of Congress Cataloging in Publication Data

Kots͡, I͡Akov Mikhaĭlovich.
The organization of voluntary movement.

(NASA TT F-16871)
Translation of Organizat͡siia͡ proizvol'nogo dvizhenii͡a.
Bibliography: p.
1. Kinesiology. 2. Neurophysiology. 3. Efferent pathways. 4. Muscle–Motility. I. Title. II. Series: United States. National Aeronautics and Space Administration. NASA technical translation; NASA TT F-16871. [DNLM: 1. Movement. 2. Neurophysiology. 3. Biomechanics. WL102 K870]
TL507.U745 no. 16871 [QP303] 629.1'08s [612'.76]
ISBN 978-1-4899-5015-4 ISBN 978-1-4899-5013-0 (eBook) 77-10938
DOI 10.1007/978-1-4899-5013-0

The original Russian text, published by Nauka Press in Moscow in 1975, has been corrected by the author for the present edition. This translation is published under an agreement with the Copyright Agency of the USSR (VAAP).

ОРГАНИЗАЦИЯ ПРОИЗВОЛЬНОГО ДВИЖЕНИЯ
Нейрофизиологические механизмы
Я. М. КОЦ

ORGANIZATSIYA PROIZVOL'NOGO DVIZHENIYA
NEIROFIZIOLOGICHESKIE MEKHANIZMY
Ya. M. Kots

Originally published by Plenum Press, New York in 1977
Softcover reprint of the hardcover 1st edition 1977

FOREWORD

Westerners commonly think that work on the nervous system in the Soviet Union derives from the school of Pavlov and his students, but in one particular area--that concerned with control of movement --the major contributions are being made by a group of scientists with rather different origins. This group, built up by the students and colleagues of Nikolas Bernstein, has its origins in mathematics, cybernetics, and kinesiology rather than in physiology. Bernstein died in 1966, but the school which he founded is active in Moscow, and Ya. M. Kots is one of its most productive members. As the work of this school has evolved, it has utilized the techniques of neurophysiology to extend the theories based on the original mathematical and kinesiological approaches. Bernstein made observations on normally functioning human subjects and inferred the cybernetic principles of operation of motor control systems by observing input and output patterns in intact individuals. Bernstein's students in mathematical and information processing institutes and schools of physical culture have extended his studies by using a variety of experimental techniques, and Kots wrote this book to summarize the current state of knowledge in this field of investigation.

Experimental results reported by Kots are derived primarily from recordings of electromyographic activity in normal human subjects. Many different experimental paradigms have been used in these experiments, but one of the major approaches has been to investigate the excitability of spinal motoneuronal pools by means of the H-reflex during periods before and during voluntary movement. The H-reflex, which involves activation of motoneuronal pools by direct electrical stimulation of afferent fibers from muscle spindle receptors, provides a test signal from which one can infer the state of excitability of spinal motoneuronal pools. Using this test object, it is possible to assess the state of spinal reflex excitability under a variety of conditions.

The book begins with a review of neurophysiological studies carried out in animal preparations, primarily by neurophysiologists outside the Soviet Union. These studies are reviewed to provide a background for the remainder of the book dealing with studies of electromyographic activity associated with volitional movements in human subjects. Kots describes a large series of remarkably ingenious experiments in which the H-reflex is used to assess the activity of the spinal gamma-motoneuron system, to study descending influences from vestibulospinal systems during movement and rest, to study interlimb reflex influences, and to investigate a whole range of phenomena associated with volitional movement.

One of the most striking overall impressions which emerges from a reading of Kots' volume is of the clear thinking and strong conceptual background with which the experimental observations are reported and brought together. The tools available to Kots and his colleagues were simple, but he has used the findings resulting from simple procedures in an extraordinarily effective way. He has brought together much of the modern neurophysiological work from outside the Soviet Union with the very extensive studies of the Moscow group. This weaving together of these two approaches makes an extremely valuable book for those interested in the organization of movement.

Edward V. Evarts

CONTENTS

INTRODUCTION

Until recently, most information on the physiology of movement was obtained from studies in two principal areas: biomechanics (psychophysiology) and neurophysiology. Biomechanical studies can provide a more or less detailed description of the phenomonology of movement, from which are derived hypotheses on the organization of a motor control system that would be able to effect the control of movements with tangible biomechanical characteristics. The most productive work in this field has been done by N. A. Bernstein (1935, 1947, 1966). On the basis of his biomechanical studies, he specified the requirements that would have to be met by a central system for the control of movement, and he formulated a number of hypotheses regarding the functional structure, as well as some general principles, of the activity of this system.

Bernstein took note of the fact that movements are accomplished by means of active muscular (internal) forces and both inertial and reactive (external) forces. The impossibility of an a priori calculation of the effect of the numerous factors involved led Bernstein to postulate a series of sensory corrections as required by the mechanism for the performance of a complete movement (Bernstein, 1947). In addition to this series of corrections based on the progress of the movement (post factum control), Bernstein believed that an essential component of the mechanism of motor control was the occurrence of a preliminary tuning of the excitability of all the sensory and motor elements involved before the onset of the movement and accomplished by the CNS in accordance with the intended motor sequence (ante factum control).

Bernstein developed the theory of a hierarchical multilevel organization of the system for the control of voluntary movements (Bernstein, 1947). According to Bernstein, the performance of different motor tasks involves the formation of different "multilevel structures," each headed by its own primary level. The primary level for a given class of movements is the lowest level on which the most fundamental--yet decisive--corrections can be accomplished. "Under its control, a number of background levels also play a role in the performance of the movement; these levels supply the background or technical components of the movement: tone, innervation and denervation, reciprocal inhibition, complex synergies, etc." (Bernstein, 1947, p. 36). The higher the primary level for the control of a movement, the higher the degree of consciousness and the more voluntary the movement.

Bernstein felt that one of the most important problems in the study of the neural mechanisms of motor control was the discovery of the nature of the interaction among different secondary subsystems (levels) of motor control. In particular, he wrote: "A spinal system consists of as many as five independent centrifugal pathways (pyramidal, rubrospinal, vestibulospinal, and two tectospinal). In the brain, we find an enormous mass of nuclei, and indications are (usually by pathology) that they play a vital role in the completion of a motor act. So far, all attempts to specifically characterize their activity have of necessity been limited to very general language and hypothetically figurative characteristics, but the fact of their activity in the synthesis of movement is not to be denied" (Bernstein, 1935, p. 9).

Bernstein's ideas regarding the principles of the function of the motor control system were widely developed in the Soviet Union (Gel'fand et al., 1961, 1962, 1966; Gel'fand and Tsetlin, 1962, 1966; Pyatetskii-Shapiro and Shik, 1964; Gurfinkel' et al., 1965; Aizerman and Andreeva, 1968, 1970; Rokotova et al., 1971).

The study of neurophysiology included: (1) the study of motor disturbances as a result of pathological or experimental injury to various nerve structures; (2) the study of the topography and nature of movements produced by the reflex path or the electrical stimulation of various supraspinal centers and conducting pathways; (3) the electrophysiological analysis of neuronal mechanisms of the central motor system. These latter studies are almost always performed with immobilized animals. Such neurophysiological studies make it possible to determine the neural structures and mechanisms that may be involved in the control of movements. By themselves, however, these experiments do not reveal which nerve centers and pathways effect the control of natural movements, or how this control is accomplished.

Evidently, these fundamental questions on the neurophysiology of movement cannot be answered without the large-scale development of yet another field of study in the physiology of movement that would combine the biomecnanical and neurophysiological approaches: the study of the participation of different nerve structures in natural movements and of the neuronal mechanisms of this participation. In this regard, studies performed in recent years in which electrophysiological methods were employed under conditions of natural movement appear to be very promising. About ten years ago, researchers began to perform microelectrode studies of neuronal mechanisms of the control of conditioned reflex (voluntary) movements in unanesthetized, freely moving animals (Chapter 1). Researchers also began to employ electrophysiological methods in the study of the neuronal activity of the cortical and subcortical structures of the brain in man during the performance of voluntary movements (Bekhtereva et al., 1967; Bekhtereva, 1971). Since 1962, studies have been made of the spinal neuronal mechanisms of voluntary motor control in man.

This book will attempt to bring together the preliminary results (including some obtained by the author) of studies on the neurophysiological mechanisms of motor control under conditions of natural motor activity in man and animals. From a neurophysiological standpoint, the organization of voluntary movement involves all processes in the CNS that precede and determine the accomplishment of a required movement or a portion of it. This book describes the methodology and results of studies of the supraspinal and spinal neuronal mechanisms governing the organization of voluntary movement.

The basic approach of our investigation of the spinal neuronal mechanisms of the organization of voluntary movement in man has been to test the state of the spinal cord during the latent period and at the onset of voluntary movement. Changes in the state of the spinal motor system occurring during the latent period of voluntary movement are the result of supraspinal (cortical, etc.) actions. Thus, by recording these changes, it is possible to identify those spinal neuronal mechanisms that are employed by the supraspinal motor systems for the organization of voluntary movement.

In our investigations, a study was made of the state of excitability of the motoneuron pool of future agonists and antagonists, and of the interneuronal spinal systems connected to them. Various approaches were developed with this goal in mind (Chapter 2). As a result of the investigations, we succeeded in obtaining preliminary data on the spinal mechanisms of the organization of voluntary movement in man (Chapters 3 and 4). In addition, studies of the nature of spinal reflex reorganization during voluntary movement caused by reflex supraspinal effects--vestibulospinal and descending interlimb effect (Chapter 5)--as well as physiological studies of neurological patients and an analysis of experimental data have enabled us to form several hypotheses on the descending systems involved in the organization of voluntary movement and on the spinal mechanisms of their activity.

We were aided in the experimental studies described in this book by Master of Physical-Material Sciences V. I. Krinskii, postgraduate students of the Department of Physiology of the State

Central Institute of Physical Culture V. I. Zhukov, A. A. Zaitsev, V. A. Mart'yanov, A. V. Syrovegin, junior fellow at the Institute for the Advancement of Physicians L. A. Makarova, and graduate student of Moscow State University V. A. Ivlev, all of whom worked in cooperation with and under the direction of the author. The author was greatly assisted in the formulation and development of his approach by correspondence with N. A. Bernstein, M. B. Berkenblit, I. M. Gel'fand, V. S. Gurfinkel', I. A. Keder-Stepanov, V. I. Krinskii, I. M. Rodionov, V. A. Safronov, F. V. Severin, V. S. Farfel', A. G. Feldmann, M. L. Tseitlin, L. M. Chailakhyan, and M. L. Shik, as well as by a discussion of the work with M. S. Zalkind, I. B. Kozlovska, and I. N. Baranov-Krylov. Valuable comments on the manuscript were provided by Academician P. G. Kostyuk, associate member of the Academy of Medical Sciences of the USSR Prof. G. N. Kryzhanovskii, and Prof. N. A. Rokotova. The author is deeply grateful to each of these comrades for his or her assistance.

Abbreviations

ATM - anterior tibial muscle
EPSP - excitatory postsynaptic potential
GM - gastrocnemius muscle (medial and lateral heads and Sol)
IPSP - inhibitory postsynaptic potential
MG - medial gastrocnemius muscle
MP - motor potential
MU - motor unit
PN - peroneal nerve
PTN - pyramidal tract neuron
PTP - posttetanic potentiation
SRIA - spinal reciprocal inhibitory apparatus
Sol - soleus muscle
TN - tibial nerve
UA - "ulnar addition"
UN - ulnar nerve
VA - "vestibular addition"

CHAPTER 1

CEREBRAL NEURONAL MECHANISMS OF THE ORGANIZATION OF VOLUNTARY MOVEMENT

Initial data on cerebral processes accompanying the organization and performance of voluntary movements were obtained by means of EEG recordings. It was discovered that the initiation of a voluntary movement is accompanied by a depression of the Rolandic alpha-rhythm, which is further diminished with continuation of the movement or with steady exertion (Spielberg, 1941; Roitback and Dedabrishvili, 1958; Dobronravova, 1963; for a survey of earlier studies, see Paillard, 1960). Studying EEG changes preceding a voluntary movement, Walter (1964) discovered a stable increase in negativity during the period between the ready signal and the signal to perform the voluntary movement. This negativity was termed the "E-wave" or the "expectation wave." It is most clearly expressed in the associative regions of the cortex, and is interrupted during the movement. The shorter the reaction time for a simple motor act, the greater the amplitude of the E-wave.

Later, German researchers (Kornhüber and Deecke, 1964, 1965) and then Americans (Gilden et al., 1966; Vaughan et al., 1968, 1970), employing modern techniques of electronic storage, averaging, and retroanalysis, revealed a gradual, incremental surface-negative shift with an amplitude of 10-15 μV that begins from 1 to 1.5 sec prior to any voluntary movement made either in response to an external signal or without a signal (spontaneous). This potential was termed the "readiness potential." It exhibits a bilateral distribution over a wide area of the cortex and is apparently analogous to Walter's E-wave. The readiness potential appears approximately 850 msec prior to a "spontaneous" voluntary movement (of a finger) and is at first bilaterally symmetrical over the pre- and postcentral areas, with a maximum over the vertex (Deecke et al., 1969). A series of tests revealed the dependence of the readiness potential on the degree of attention, instruction, motivation, expectancy, and resolve, as well as a relationship between the amplitude of the readiness potential and the reaction time (Kukinova and Ivanova, 1974). As the subject nears initiation of the movement, there is a further increase of the negativity over the contralateral precentral area (Kornhüber and Deecke, 1965) with a somatotopic localization corresponding to the projection zone of the moving member in the motor cortex (Vaughan and Costa, 1968). The latter authors identify this late part of the negative potential with the early part of the motor potential and regard it as a cortical correlate of preparatory motor tuning.

The readiness potential is followed first by the onset of a bilaterally symmetrical positivity that appears 80-150 msec prior to EMG of the agonists and that varies considerably from one subject to another; this potential was termed the "premotor potential" (Deecke et al., 1969). Next, a negative potential, known as the "motor potential" (MP) (Kornhüber and Deecke, 1964, 1965; Deecke et al., 1969), appears approximately 40-60 msec prior to EMG of the agonists and exhibits a peak over the contralateral projection motor region of the cortex (Vaughan and Costa, 1968; Vaughan et al., 1968; Deecke et al., 1969). Thus, based on its spatial distribution, the MP is closely associated with that area of the motor cortex that must be active for the performance of a given movement (Vaughan et al., 1968). It was shown in a study by Nebylitsyn and Bazilevich (1970) that MP amplitude is a function of the force exerted during voluntary movement.

Vaughan and Costa (1968) used summation to determine the individual EEG components for conditions of stimulation during rest and during the preparation and performance of a voluntary movement. An electronic computer was used to detect both the EEG control response to sensory stimulation during rest and the response to the same stimulus prior to and during voluntary muscular contraction. The authors regarded the difference obtained as corresponding to the MP associated with corticospinal discharge. According to their data, during a simple motor reaction, the cortical sensorimotor central delay ranges from 20 to 60 msec, while the efferent period (from the appearance of MP to EMG) ranges from 30 to 60 msec for wrist muscles and 60 to 120 msec for foot muscles. In a reaction involving a choice of two signals, both the central delay and the efferent period are lengthened. Vaughan et al. (1970) noted that the efferent period is substantially lengthened when the movement is spontaneous rather than in response to a signal.

The entire complex of EEG changes that precedes the onset of a voluntary movement is undoubtedly a manifestation of the functioning of different neural mechanisms (structures) that contribute to the organization of voluntary movement. Unfortunately, the nature of the various EEG components remains obscure.

Our knowledge of the cerebral neuronal mechanisms of the organization and regulation of voluntary movement has expanded greatly since the initial use of electrophysiological methods in the study of animals performing natural movements. The use of such methods made it possible to study the activity of motor cortex neurons during spontaneous and conditioned movements. The first such studies in which chronically implanted microelectrodes were used to record the activity of different cortical neurons during avoidance conditioning of monkeys were performed by Jasper et al. (1962). This approach was much refined by Evarts, who used microelectrodes to record the activity of the neurons of different cerebral structures both prior to and during conditioned reflex movements in monkeys.

1. THE MOTOR CORTEX

Evarts began his investigations with a study of the activity of motor cortex neurons, identifying them as pyramidal tract neurons (PTNs) by antidromic stimulation of bulbar pyramids via chronically implanted electrodes. During this series of investigations (Evarts, 1965), it was found that pyramidal tract neurons located in the motor cortex could be separated into two groups according to axonal conduction velocity: neurons with higher conduction velocities were active during movement of the contralateral limb, whereas neurons with lower conduction velocities tended to have constant tonic discharge with frequency variation during movement. In a number of cases in which simultaneous recordings were made from two pyramidal neurons of the motor cortex, Evarts discovered that their behavior was often reciprocal during movement (Evarts, 1964).

After the development of a standard conditioned reflex method, a detailed study of the relationship between the activity of motor cortex neurons and the conditioned reflex movement was begun. Evarts (1966) trained monkeys to perform a simple movement--flexion or extension of the wrist--in response to a light flash. The monkey received a liquid reward if the latent period from the light flash to initiation of the movement did not exceed 350 msec. Microelectrode recordings were made of the activity of separate motor cortex neurons in the projection zone of the hand, and simultaneous EMG recordings were made for contralateral wrist movements. In about one-third of the motor cortex neurons, an increase or decrease in discharge frequency was observed prior to EMG of the contralateral (and occasionally the ipsilateral) hand muscles. The neurons began to discharge with a minimum latency approximately 100 msec after the light flash, and this discharge preceded the onset of the agonist EMG volley by about 100 msec. A clear relationship was established between neuron response latency (change in frequency or appearance of impulse discharge) and EMG latency (reaction time).

In addition, Evarts found a strong positive correlation between the performance of a conditioned reflex movement and the change in the discharge of a pyramidal neuron: if there was no movement during the extinguishing of the conditioned reflex, there was no neuron response. Hence, neuron discharge appears to be not merely a response to the light flash, but to be associated in some way with the performance of the movement. In cases when neuron activity changes prior to initiation of the movement, this activity cannot be associated with afferent input arising from the moving limb. These observations provided strong support for the hypothesis that there is a causal relationship between the initiation of a movement and the activity of motor cortex PTNs. Some PTNs sharply reduce their discharge frequency prior to EMG, but with these neurons, no relationship is observed between the nature of the frequency change (increase or decrease) and the direction of movement (extension or flexion of the wrist). Other PTNs begin to discharge only after the start of the EMG; their activity may thus be the result of peripheral feedback, or it may be associated with incipient muscular contraction in the conditioned response sequence.

Very important data were obtained from experiments performed by Evarts (1967) in which the activities of two or more neurons were simultaneously recorded during the performance of voluntary movements. Evarts proposed that neighboring PTNs acting to _excite_ motoneurons innervating the same muscle should have positively correlated discharge in much the same way as two motoneurons innervating the same muscle; such a pair of PTNs should not show reciprocal activity, but should have a fixed relationship (i.e., the relationship should not vary with variations in the movement and underlying pattern of muscular contraction). In contrast, Evarts proposed that two PTNs controlling different muscles should have a _plastic_ relationship (i.e., should show synchronous discharge for some movements and reciprocal activity for others), just like the motoneurons through which they act.

Evarts' studies showed that the great majority of PTN pairs have plastic connections, and are thus analogous to spinal motoneuron pairs that innervate two different muscles. Not a single pair of PTNs was found to have fixed reciprocal connections, and only a few pairs exhibited a fixed positive correlation, i.e., the type of connection that can be observed in two spinal alpha-motoneurons that innervate the same muscle (see Chapter 4, Section 1).

Because the important or even decisive role of the pyramidal system in the initiation of a voluntary movement was a widely accepted concept, Evarts' data on the activity of the cortical pyramidal neurons preceding the onset of a voluntary (conditioned reflex) movement were regarded as entirely natural. However, physiologists were surprised by such a substantial delay between pyramidal neuron discharge and the onset of impulse activity by spinal motoneurons (EMG), this delay amounting to approximately 100 msec according to Evarts' preliminary data. In this respect, Evarts' findings were supported in studies published by Luschei et al. (1968, 1971), and by Porter et al. (Porter et al., 1971; Porter and Muir, 1971).

Recording the activity of individual neurons located at various depths in the motor cortex of monkeys during the performance of a conditioned reflex movement (pressing a button for a food reward), Luschei et al. (1968) found that some of the neurons exhibit an increase or decrease in discharge frequency prior to movement. Judging by the recordings presented in the article, the change in the discharge of such neurons begins more than 100 msec before the button is pressed. In another study (Luschei et al., 1971), traces were made of the activity of neurons of the precentral motor region (Field 6) in the motor projection zone of facial muscles during movement in monkeys conditioned to flex their jaw (bite) in response to a light flash with a food reward. Of 161 neurons, 37 exhibited an increase and 17 a decrease in discharge frequency, an average of 50 msec prior to movement. During masticatory movements, these "early" neurons did not always exhibit a relationship to the jaw movements. The "early" neurons did not change their activity in accordance with movements of the tongue or mimicking movements, and did not react to sensory stimulation. The other neurons studied changed their frequency only after the onset of the movement, and some were activated by sensory stimulation applied in the area of the mouth.

In discussing such a substantial delay between "early" neuron discharge and movement, Luschei and his co-authors do not discount the possibility that these neurons are involved in the modulation of somatosensory input that apparently occurs prior to movement (see Section 2). If the "early" neurons do have a motor function, their control of motoneurons is indirect and has to do with the recruitment of subcortical neuronal systems.

In experiments conducted by Porter et al. (1971), rhesus monkeys performed conditioned reflex movements for a food reward without presentation of a signal. The traces from the pyramidal and nonpyramidal neurons of the precentral cortex also revealed a positive correlation between their discharges and the beginning (or end) of the EMG of shoulder and forearm muscles. The average delay between the first impulse in the discharge of such neurons and the initiation of movement was about 80 msec, which is close to the figure obtained by Evarts (1966) and Luschei et al. (1968). In discussing such a long delay between pyramidal neuronal discharge and muscular contraction, Porter and colleagues point out that this fact cast some doubt on the role of the pyramidal system in the rapid initiation of movements (Wiesendanger, 1969). Porter et al. admit that some temporal facilitation is required even for the most direct (corticomotoneuronal) projections, this facilitation being provided by the rhythmic discharge of the pyramidal cortical neurons, which is an effective initiator of muscular contraction. Using stimulation to evoke a high-frequency discharge of the pyramidal tract (up to 500 imp/sec), Porter et al. were able to shorten the delay between cortical discharge and muscular contraction to 35 msec. Since the natural discharge frequency in corticomotoneuron fibers is usually less than 100/sec (Evarts, 1966, 1968), the delay may extend to several dozen milliseconds.

Porter (1972) turned his attention to the variability of the intervals between impulses in a cortical neuron volley. The inter-impulse intervals associated with the performance of a given movement often diminish and later increase within the limits of the volley. With an average discharge frequency less than 100/sec, some intervals are as brief as 3 msec, the shortest intervals usually occurring in the middle of a neuron volley. On the basis of such observations, it was hypothesized that the delay between the onset of pyramidal neuron activity and spinal motoneuron activity is determined to a significant degree not only by the average frequency, but also by the temporal pattern of a pyramidal neuron volley. Porter and Muir (1971) showed that the facilitatory effects in spinal motoneurons are, in fact, strongly dependent on the time course of a pyramidal tract volley. For equal stimulation frequencies of the corticospinal pathways, the times at which the maximum excitatory effect in spinal motoneurons is achieved for different temporal patterns of stimulation can vary by as much as 50-75 msec. The time course of a volley can also cause significant differences (up to 50%) in the magnitude of the evoked depolarization of the motoneuronal membranes. On the basis of their data, Porter and Muir concluded that one of the functions of a pyramidal neuron discharge is to determine the timing of muscular contraction during cortically controlled movements. The latency of a muscular contraction may possibly be coded by the pattern of intervals between nerve impulses in a volley produced by PTNs.

This hypothesis was confirmed in the next study by Porter (1972), in which the relationship between the time course of a volley of cortical cell impulses and the onset of a spontaneous stereotype movement in monkeys was investigated. A significant relationship was found between the statistical characteristics of a volley from individual cortical cells and latency of movement, which can be determined by the interval following the first spike in the volley. A very high correlation was established between the length of the delay from the smallest interval in the volley to the onset of movement, and from the first impulse in the volley to the onset of movement.

The principal question seems to be that of a possible temporal connection between cortical neuron discharge and the onset of voluntary movement, because its solution is particularly related to the question of a causal relationship between the discharge of motor cortex neurons and the initiation of voluntary movement. In all the experiments of Evarts and his successors, however, movements that could be performed by animals involved the activity of a number of muscles. Therefore, such experiments do not indicate how these muscles are associated with the recorded cell of the motor cortex.

To determine the extent of the relationship between the activity of a precentral cell and the contraction of specific limb muscles, Fetz and Finnocchio (1972) trained a monkey to isometrically contract only one of four muscles at a time. The monkey's behavior was reinforced whenever the weighted sum of the activities (EMG) of the muscles and cell exceeded a criterion value. For example, while the monkey was being trained to contract a single muscle, the weighting factor of this muscle was such that during its activation, a preselected value was attained in an integrator, which in turn activated a food-reward relay. After 6 weeks' training, the monkey began to contract each of the four muscles individually. The authors recorded 9 precentral neurons (6 of which were pyramidal) during the individual contraction of each of the four hand muscles. Of these neurons, 3 modified their activity in connection with the contraction of only one or two of the four muscles, 4 exhibited some connection with each of the four muscles (2 of the neurons produced an identical volley in connection with all four muscles), and the remaining 2 neurons exhibited no correlation with any of the muscles. Under conditions of isometric muscular contraction, the activity mode of neurons associated with the activity of two antagonists of the same joint was more frequently identical (6 cases) or indistinguishable (5 cases) than reciprocal (3 cases). Two cells became active prior to the isometric contraction of both the biceps and triceps, but were reciprocally associated with active flexion and extension in the elbow joint, becoming active prior to one movement and inactive prior to the other. In 6 cases in which cellular activity correlated with a particular muscle, the burst of cellular activity was reinforced by the simultaneous depression of the activity of all muscles. In every case, the monkey easily learned to excite a cell without significant correlation with EMG activity. Reverse dissociation (muscular activity and depression of cellular activity) was tested twice. In one documented case, the response pattern was shifted toward the reinforced situation, but cellular activity was never completely depressed. These interesting experiments should be supplemented by similar tests.

In another of his studies, Evarts (1968) attempted to determine which parameters of voluntary movement--amplitude, speed, or force--the pattern of the impulse activity of pyramidal neurons is correlated with. In one experiment, the amplitude of a monkey's wrist movement was limited by means of special apparatus (to 30°), and the force that had to be exerted was varied by means of different resistances (weights) that had to be overcome during either a flexion or extension movement. Owing to its training, the monkey could perform a movement with a constant amplitude and could flex or extend its wrist by the contraction and partial relaxation of that group of muscles working against the weight. Evarts discovered that in most cases, PTNs discharged only during activation of a particular muscle group (flexors or extensors), independent of the direction of movement (extension or flexion). For example, when flexion was opposed by a moderate weight such that both flexion and extension of the wrist were accomplished by a change in the degree of contraction of flexors only, the neuron in question discharged for both directions of wrist movement, whereas the same neuron became completely inactive when the wrist movement was performed by the contraction of extensors (weight opposing extension of the wrist). The reverse was observed in the case of pyramidal neurons the activity of which correlated with extensor contraction.

Thus, in most cases, the excitation of PTNs that control wrist movement is associated with the magnitude and rate of change of the force exerted by the muscle controlled. It is significant that under natural conditions of movement, the cortical neurons of the pyramidal tract are associated with both the contraction of flexors and the contraction of extensors, unlike the effects of electrical stimulation of the pyramidal tract, which lead primarily to the facilitation of flexor motoneurons (see Chapter 3, Section 2).

To investigate further the connection between neuronal and muscular activity, an experimental situation was created in which a monkey had to exert a constant force in order to keep its wrist in the same position and obtain a reward (Evarts, 1969). In different experiments, this was achieved by the activity of either flexors or extensors working against a weight. The maintaining of a constant wrist position was accompanied by slight, short-term fluctuations of force. These slight increments and decrements (first derivative of the force) were measured independently of the constant (background) magnitude of the force. It was found that the activity of most pyramidal neurons is more closely related to decrements and increments in force (first and second derivative) than to the force itself. Moreover, the changes in the discharge of these neurons related to slight fluctuations of force are almost independent of the magnitude of the constant force. These findings indicate that such pyramidal neurons are involved in motor control over a wide range of forces, since their discharge apparently controls only changes (increments or decrements) in force.

For some neurons, the range of forces in which a relationship is observed between the force derivative and discharge frequency is limited, e.g., under conditions in which a constant extension force is exerted against a weight opposing extension of the wrist. Such neurons no longer reflect changes in force and its first derivative when the weight opposes flexion of the wrist, i.e., under conditions of relatively strong flexor activity and practically no extensor activity. For other neurons, the operative region of constant force in which discharge frequency is related to force and its derivative is within the range of flexor forces: the activity of such neurons reflects changes in the first derivative when flexion is opposed, but not extension. The third type of neuron is active for all weights employed (from 400 g for flexion opposition, up to 400 g for extension); their discharge is associated with the force derivative, regardless of the magnitude of the force; i.e., such units are associated with the force derivative, regardless of the constant level of force onto which the fluctuations are superimposed. These findings by Evarts led to the conclusion that the force of muscular contraction is the "primary" output variable of cortical pyramidal control. In this case, the speed and amplitude of the movement must be "secondary" variables, the magnitude of which is a consequence of the controlled variable--the force of muscular contraction (Evarts, 1967).

In a discussion of Evarts' data, V. B. Brooks and Stoney (1971) indicate that the force and speed of movement are determined by the same parameters--the number of active motor units, the degree of their synchronization, and the relative forces of the antagonistic muscles. Because there is an intercorrelation between the forces exerted for the execution of a movement and the speed of the movement performed by a monkey (Humphrey et al., 1970), a distinction can scarcely be made between the control of muscular contraction force and the speed of movement. In an experimental situation similar to that employed by Evarts, V. B. Brooks et al. (1972) studied the relationship between motor cortex cell discharge and the speed of hand movement in monkeys during the performance of alternating movements in the wrist joint against various modes of resistance with an amplitude of about 30° (without mechanical motion arrestors), a voluntary speed of movement, and postural fixation within a certain period of time (no less than 1 sec). In accordance with data obtained by other authors, the activity of some cortical cells was found to be related to the external task (Evarts, 1968) and the speed of movement (Humphrey et al., 1970; Humphrey, 1972). The normal linear relationship between the phasic discharge frequency of a cortical cell and the maximum speed of movement was violated when the dentate nucleus of the cerebellum was cooled.

Using a multielectrode to simultaneously record the activity of several (3-8) motor cortex neurons and employing special mathematical methods of analysis, Humphrey et al. (1970) showed that data on a simple voluntary movement (analogous to the movement in Evarts' experiments) are provided not only by the pattern of the discharge of individual cortical neurons, but also by the nature of the temporal interrelationship of the activity of the neurons. In experiments conducted by these authors, monkeys were trained to move a vertical rod from side to side, flexing and extending the wrist in opposition to a variable resistance (weight). The pressure exerted on the rod and its displacement (angle in radiocarpal joint) were recorded. The correlation between neuronal discharge frequency and movement parameters was highest when the latter were shifted back 100 msec with respect to the former. Although each of the cortical cells changed its discharge frequency during a specific phase of movement, the correlations between these changes and movement parameters were not very high and were inconsistent. In addition, by "weighing" and summating the discharge frequencies of the concurrently recorded motor cortex neurons, the authors were able to predict with a high degree of accuracy the time course of force and speed parameters. This accuracy improved as the number of concurrently recorded neurons was increased.

Like Evarts (1966), Humphrey et al. (1970) determined that each of the simultaneously recorded cortical cells is related to the first derivative of the force of muscular contraction, although the moment during the movement at which changes in discharge frequency exhibit a maximum was found to vary from one cell to another. In view of this circumstance, it is not surprising that the combined activity of all the simultaneously recorded cells usually correlates better with temporal changes in force and other motor characteristics (displacement and speed) than with the first derivative of force. The correlation factor between the combined activity of all the neurons and the speed of movement was only slightly lower (0.7) than for the force parameters, and did not vary as the limb was subjected to different loads. In a study by Humphrey et al. (Humphrey et al., 1970; Humphrey, 1972), it was shown that the amplitude of conditioned wrist movements in monkeys can be reliably predicted (correlation factor 0.56) in cases when the discharge of 3-8 neurons is being simultaneously analyzed. This correlation shows, however, that only 31% of the variance between the predicted and observed amplitudes of limb displacement can be accounted for by the time course of a pyramidal tract discharge, whereas 71% of the variance can be explained in the case of force. Thus, it is still impossible to provide a simple answer to the question "Which motor parameters exhibit the highest correlation with, and are thus possibly controlled by, the activity of motor cortex neurons?" We know only that motor control is fundamentally influenced not only by the pattern of a cortical neuron volley, but also by the temporal relationships between neuronal discharges.

2. THE SOMATOSENSORY (POSTCENTRAL) CORTEX AND SOMATIC AFFERENT PATHWAYS

Experiments in which the activity of individual neurons of different cerebral structures was recorded before the performance of a voluntary (conditioned) movement have made it clear that the initiation of a voluntary movement is associated with complex reorganizations occurring not only in the motor cortex, but also in many other (including afferent) neuronal systems of the brain.

It is known that the postcentral somatosensory cortex in monkeys has efferent projections (including projections into the pyramidal tract) that do not depend on integration in the primary or supplementary motor areas (Woolsey et al., 1953; Lassek, 1954; Woolsey,

1958; Russell and Meyer, 1961). By contrast, the precentral motor cortex receives a variety of sensory inputs, including some from the associative areas of the cortex, as well as "direct" afferent projections independent of the somatosensory areas of the cortex (Gardner and Morin, 1953; Malis et al., 1953; Byuzer and Ember, 1964; Voronin and Tanenholz, 1967). Thus, the pre- as well as postcentral somatic cortex in primates (and homologous areas in other mammals) may be involved in motor as well as sensory processes (Woolsey, 1958; Batuev, 1970).

In a study published in 1972, Evarts (1972) attempted to determine whether there is a change in the activity of neurons of the postcentral sensory region, and if so, whether it occurs prior to movement or only after the onset of movement as a result of sensory feedback from the periphery. He therefore recorded the activity of neurons of the postcentral area of the cortex in monkeys conditioned to perform a rapid flexion or extension of the wrist in response to a light stimulus. Evarts determined that the postcentral (nonpyramidal) neurons become active later than the precentral (pyramidal and nonpyramidal) neurons. In most cases, the precentral neurons begin firing approximately 60-40 msec before the initiation of pressure against a rotating handle (i.e., 30-10 msec prior to voluntary EMG of the agonist), and the postcentral neurons, 0-20 msec after initiation of pressure (i.e., 30-50 msec after start of EMG of agonist). Since most postcentral neurons discharge only after contraction has begun, their discharges may be the result of sensory feedback. The time of the earliest postcentral activity (60-40 msec prior to pressure) coincides with the period of reduced antagonist activity preceding agonist activity (the Wachholder-Hufschmidt phenomenon). Hence, even early postcentral neuronal activity may be the result of feedback.

Evarts argues that this feedback is basically the result of peripheral action, but he does not discount the possibility that discharges in some postcentral neurons may actually begin prior to the action of peripheral feedback. He bases this belief on the theory of corollary discharge (Sperry, 1950; Teuber, 1966), whereby impulses must leave the motor area and enter the sensory area during the initiation of voluntary movement, the sensory area thereby being "warned" of the impending movement before the arrival of peripheral feedback (see also Kots, 1966; Kots and Naidin, 1966). This mechanism may make use of abundant connections of the precentral motor cortex with the postcentral sensory cortex (Jones and Powell, 1969) that enable the activity in the precentral neurons prior to movement to modify the activity of neurons in the postcentral cortex even before the arrival of sensory feedback after the movement has begun.

In addition to inputs from the motor cortex, the postcentral gyrus also receives afferent input from the cerebellar dentate nucleus through the ventral posterolateral nucleus of the thalamus, which may also be excited via the classic lemniscus pathway (Bava et al., 1966; Nakagoshi, 1966). Since neurons of the dentate nucleus become active prior to movement (Thach, 1970a,b; see below), the postcentral neurons may be excited prior to movement through a cerebellothalamocortical input. These "nonsensory" inputs to the sensory cortex may not be able to evoke a discharge by the all-or-none rule, but they can serve to modify the excitability of postcentral neurons as well as alter the all-or-none discharge evoked by sensory feedback (Evarts, 1972).

In addition to changes in the sensory cortex, fundamental changes occur in the ascending somatic sensory transmission system itself during the organization and performance of a voluntary movement. A large volume of anatomic and electrophysiological data obtained from experiments with immobilized animals has demonstrated the possibility of centrifugal effects on the transmission of ascending afferent impulses. These effects, particularly pyramidal effects, may occur at different levels of sensory transmission (for surveys, see Wiesendanger, 1969; Ghez and Lenzi, 1971), beginning with neurons of the dorsal horns of the spinal cord (Wall, 1967). It was shown in a series of studies that the transmission of an afferent volley through the nuclei of dorsal columns and other relay points for the transmission of somatic information to the medial lemnisci can be modified by very diverse peripheral and central effects (Norton, 1969) from the sensorimotor cortex, cerebellum, nonspecific thalamic nuclei, and other central structures. Such effects have also been observed under physiologic conditions--during some phases of sleep, for example (Carli et al., 1967), or during spontaneous movement in unrestrained rats (Ainsworth et al., 1969; O'Keefe and Gaffan, 1971). In particular, it was found that the depression of afferent signal transmission through the cuneate nucleus that occurs during the rapid-eye-movement phase of sleep coincides with an increase in the activity of pyramidal tract neurons (Evarts, 1964).

In studies by Ghez and Lenzi (1970, 1971), it was determined that for a long period preceding and during a conditioned movement, there is a depression of the transmission of somatic (cutaneous) afferentation through the lemniscal system. These authors trained a cat to raise its forepaw, press a lever, and lower its paw to the floor in response to an audible signal. During the various phases of movement, they recorded the responses in the contralateral medial lemniscus to electrical stimulation of the superficial radial nerve. More than 100 msec prior to the paw movement, the amplitude of the evoked lemniscal response begins to decay; depression of the evoked response attains a maximum approximately 100 msec prior to movement; and the amplitude of the response remains low during both raising and lowering of the forepaw. The latter fact is also described by Coulter and Thies (1971). Control experiments showed that depression of the evoked lemniscal response is not related to the influence of the conditioned signal itself or to a reduction of the afferent volley in the stimulated nerve. Since depression of transmission occurs prior to movement, it is evidently associated with descending effects on afferent relay points in the lemniscal system.

In a later study, Ghez and Pisa (1972) determined that the degree of depression of somatosensory transmission (lemniscal response) is directly proportional to the logarithm of the speed of movement of the forepaw. On the other hand, the force exerted during the movement does not determine the degree of depression of the lemniscal response. Wall's method involving the determination of excitability in the afferent terminals of the cuneate nucleus (by the magnitude of the antidromic volley) was employed to discover the depolarization of these fibers during the performance of both flexion and extension movements. This phenomenon preceded contraction of the flexors (raising of the forepaw) by 100-200 msec. The degree of facilitation of the antidromic volley (depolarization) of the afferent fibers of the cuneate nucleus exhibited a statistically high correlation with the logarithm of the speed of movement, given no significant differences between flexion and extension. There was a weak but statistically significant depression of the initial alpha-wave of the orthodromic potential in the medial lemniscus on stimulation of the cuneate nucleus during voluntary flexion or extension of the hind limb. The appearance of this depression preceded contraction of the flexor. Thus, the data obtained indicated that the transmission of a volley from cutaneous receptors to the medial lemniscus may be depressed prior to and during voluntary movement as a result of pre- and postsynaptic inhibition in the cuneate nucleus.

It is known that the pre- and postsynaptic inhibition of cuneate nucleus neurons can be evoked by both central and peripheral effects (Andersen et al., 1964). The reduction in lemniscal potentials, primary afferent depolarization, and the depression of the alpha-cuneothalamic response prior to movement are easily explained by central effects on the cuneate nucleus. During motor performance, it is difficult to distinguish the relative participation of central and peripheral factors in the observed changes. The absence of a clear connection between inhibition of lemniscal transmission and the force exerted during movement is interesting in the light of the high correlation (described above) between pyramidal discharge and the force of muscular contraction (Evarts, 1968). The speed, however, is also "represented" in the cortical discharge (Humphrey, 1972). At the same time, it should be noted that other central structures besides the cerebral cortex can influence transmission through the cuneate nucleus (Sotgiu and Cesa-Bianchi, 1970).

There is evidence that similar pre- and postsynaptic inhibitory actions can also influence the transmission of joint and muscular afferentation to the cuneate nucleus. For example, Carli et al. (1967) discovered that during rapid eye movements in desynchronized sleep, primary afferent polarization is observed in both cutaneous and muscular terminals to the cuneate nucleus. The observations of Coquery (1971) are of interest in this regard: the perception of an electrical stimulus applied to the human finger is diminished or completely disappears prior to and during voluntary flexion of the finger. This author also found that the cortical potential evoked in a human on electrical stimulation of the finger increases 200 msec prior to voluntary flexion and declines after the onset of the movement.

Vaughan et al. (1970) discovered that an evoked response is observed in the EEG in monkeys at the onset of a passive movement, but is absent during active voluntary movement--a conditioned extension of the wrist. This finding led the investigators to conclude that kinetic feedback to the motor cortex is blocked during the performance of a conditioned voluntary movement. In the opinion of Vaughan and his co-authors, the performance of this class of movement is made possible by the presence of a central motor program formed during conditioning that is carried out without participation of the peripheral circuit (kinetic feedback). Such a conclusion is supported by their finding that following complete uni- or bilateral upper-limb deafferentation in monkeys, the motor potential in the EEG (see Section 1) accompanying the performance of a conditioned movement did not exhibit an appreciable change; however, there is an increase in the delay from the onset of motor potential to the beginning of the voluntary movement.

The depression of transmission in the somatic afferent system that precedes the onset of a conditioned movement can be regarded as one of the mechanisms for switching off the central motor control system from that portion of peripheral information that is unessential for the performance of a given movement (or a given phase of the movement). It is significant that this switch-off occurs before the movement begins.

The role played by thalamic relay nuclei in modifying the afferent discharge before voluntary movement remains obscure. While making recordings during stereotaxic operations on patients suffering from paralysis agitans, Jasper and Bertrand (1966) discovered an increase in the activity of ventrolateral thalamic neurons prior to voluntary movement. These neurons are activated by voluntary movements of one limb, but not by passive movements of the same limb or by palpation of its muscles or tendons. The authors claim that such cells must be involved in the initiation of movement, rather than in merely reflecting sensory response to movement.

According to the data of Evarts (1970), the activity of ventrolateral thalamic neurons in monkeys changes more than 100 msec prior to the onset of a conditioned movement (flexion or extension of the wrist in response to light flash with preliminary fixation of flexed or extended wrist position), i.e., approximately when the earliest premovement changes in the activity of motor cortex neurons appear. These findings led Evarts to believe that thalamic neurons play a decisive role in controlling the activity of the motor cortex prior to movement. According to Evarts, this role of the ventrolateral thalamus in the organization of movement does not exclude its second, "classic" function as a relay for sensory feedback after movement is initiated.

3. THE SUBCORTICAL NUCLEI AND THE CEREBELLUM

Evarts' method of recording neuronal activity during the performance of a conditioned movement by an animal was also applied in determining the motor functions of the basal ganglia, particularly in elucidating their role in the control of voluntary movements (De Long, 1971a,b, 1972). According to the data of De Long (1971a,b), most pallidal neurons in monkeys increase their discharge frequency during the performance of "spontaneous" movements. In some cases, the discharge of the neurons is strongly related to specific movement (tongue, hand, or foot), though such a specific relationship is usually not observed. During conditioned movements, most neurons of the internal and external segments of the pallidum (more than 80%) exhibit phasic discharges that have a definite temporal relationship with movements of the contralateral arm. The activity of most neurons of the internal and external segments of the pallidum is specifically related to movements of only one limb, and is frequently of a different nature during different movements of the same limb. In terms of their localization in the pallidum, the neurons associated with upper- and lower-limb movements have overlapping spatial loci and are situated chiefly in the lateral portions of the internal and external segments, i.e., in those portions of the pallidum receiving the projections from the motor and somatosensory areas of the cortex (Nauta and Mehler, 1966) that pass through the corticostriopallidal pathway (Kemp and Powell, 1970).

To determine the temporal relationship between changes in the activity of basal ganglia neurons and the initiation of voluntary movement, De Long trained monkeys to perform a rapid hand movement in response to a light flash (De Long, 1972). He discovered that most neurons of both segments of the pallidum change their discharge prior to movement, this change being more closely associated with the onset of the movement than with the presentation of the light flash. The activity of neurons of the putamen is analogous to that of the pallidal neurons. De Long concluded that the basal ganglia participate in the initiation of voluntary movement (see also Denny-Brown, 1962). While making extracellular microelectrode recordings of the impulse activity of subcortical neurons during operations on paralysis agitans patients, Livanov and Raeva (1972) recorded a transient change in neuronal activity in extrapyramidal structures (pallidum, putamen) that accompanies the initiation and termination of a voluntary movement (making a fist).

That the earliest activity changes of the cells of the pallidum and putamen occur prior to the onset of a conditioned movement indicates that these basal nuclei are involved in some way in the initiation of movement. The pallidum can influence the precentral cortex prior to movement particularly through the thick pallidothalamocortical pathway (Nauta and Mehler, 1966). It would thus be important to determine whether the activity of basal ganglia neurons precedes activity in the motor cortex. The preferential relationship between the activity of pallidal and putamen neurons and contralateral movement, and the specific relationship between their activity and the direction of movement, indicate that there is a detailed "representation" of the aspects of movement at this level. However, De Long notes that the discharge of basal ganglia neurons is less closely associated with movement than is the discharge of motor cortex neurons (Evarts, 1966). In qualitative terms, the former is more similar to the discharge of neurons of the cerebellum (Thach, 1968, 1970a,b), which, like the basal ganglia, is separated from the motor "output" by several synapses.

Ideas regarding the participation of the cerebellum in the regulation of voluntary movements have been expressed for some time (Herrick, 1924; Holmes, 1930; Orbeli, 1934; Dow and Moruzzi, 1958). According to modern theories (Eccles, J. C., et al., 1967; Evarts and Thach, 1969), the cerebellum integrates information relayed to it through various nerve pathways, and the resulting activity of the cerebellum is (1) transmitted via brain stem structures into the spinal cord to the motoneurons, in which case the cerebellum is directly involved in the control of movement; or (2) carried by various relay nuclei into the cerebral cortex, in which case the cerebellum is involved in the control of movements carried out by the higher motor centers. In particular, J. C. Eccles et al. (1972) postulate that the cerebellum provides a continuous flow of corrections that are incorporated by the cerebrum into its pyramidal output, thereby providing a sensitive monitoring of the performance of skilled movements.

Herrick and Holmes formed two different hypotheses regarding the participation of the cerebellum in the initiation of voluntary movement (see Evarts and Thach, 1969). According to Herrick (1924), the activity of the prefrontal cerebral cortex, which elicits activity in the precentral motor cortex, also activates the corticopontine pathway, sending commands to the cerebellum: "I am about to leap forward; get set, and cooperate in the movement." By the time the appropriate lower motor centers are activated through the pyramidal tract, the action of the cerebellum on these lower centers has taken effect, increasing the force of muscular contractions and producing a more effective movement. Holmes (1930) discovered that the onset of a voluntary movement was appreciably delayed in patients with lesions of the cerebellum. These difficulties at the onset of a voluntary movement, Holmes argues, result from damage to the cerebral mechanisms of voluntary movement, producing a delay in the initiation of corticospinal innervation. He surmised that cerebellar activity normally precedes the onset of voluntary movement and aids in its initiation. It is significant that delays in the initiation of movement due to cerebellar lesions are particularly pronounced in the chimpanzee and in man (Holmes, 1930), in whom the cerebellum has established exceptionally thick connections with the cerebral cortex.

The hypothesis offered by Herrick and Holmes received good experimental corroboration in studies performed by Thach (1968, 1970a,b, 1972). By recording the activity of individual neurons--Purkinje cells, cells of the cerebellar output nuclei (dentate and intermediate)--in monkeys during the performance of conditioned movement, Thach found that activity of the cerebellum changes prior to movement. He also showed that changes in the discharges of the neurons of the dentate and intermediate nuclei are temporally related to motor performance, but not to the conditioned signal (Thach, 1970a). The following facts tend to support this assertion. First, when an animal did not make a movement in response to the signal, there was usually no change in neuronal discharge. Second, when a movement was performed without the signal, the changes in the discharge pattern were the same as those accompanying a movement made in response to the signal. Third, changes in neuronal discharge were generally different for movement in different directions (flexion or extension). With most neurons, a change in discharge frequency occurred at different times during flexion and then during extension of the wrist. The change in neuronal discharge consisted of an increase (most often) or a decrease in discharge frequency, and could commence prior to movement (most dentate nuclear neurons and about half the interpositus nuclear neurons) or after initiation of movement. Neurons of the dentate nucleus became active earlier than neurons of the interpositus nucleus. The earliest changes in the discharge of the neurons of the cerebellar output nuclei preceded the initiation of movement (EMG of primary agonists) by several dozen milliseconds (up to 90).

Thus, Thach's experiments revealed that the neurons of the dentate and interpositus nuclei of the cerebellum may change their discharge in connection with a rapid conditioned (voluntary) movement, either prior to or during the movement. That the discharge pattern (or frequency) of these neurons is usually different during the performance of movements in different directions shows that the cerebellar nuclear output may influence the choice of direction. That many neurons of the cerebellar output nuclei (especially the dentate nucleus) modify their discharge frequency long before the initiation of movement suggests that the cerebellum participates in the organization of voluntary movement. This is in agreement with the hypothesis offered by Holmes (1930) to the effect that changes in the discharge activity of the dentate nucleus precede and initiate activity changes in the pyramidal tract.

In discussing his results, Thach (1970b) points out that motor cortex PTNs modify their discharge prior to movement with a delay that is similar to that of neurons of the dentate nucleus and that permits the latter to influence the motor cortex through the cerebellothalamocortical pathway (Yoshida et al., 1966). Another possible explanation is that changes in dentate nuclear discharge may follow the earliest pyramidal tract discharge, in which case cerebellar effects can modify the cortical output after it has begun but still prior to actual movement. This idea is close to the theory of J. C. Eccles et al. (1967) pertaining to the function of the cerebellum.

According to data obtained by Thach, some neurons of the dentate nucleus and many neurons of the intermediate nucleus modify their discharge frequency after the onset of movement. In this case, the changes may be the result of the partial accomplishment of movement and be induced by afferentation. The activity of such neurons may assist in the completion of the movement, i.e., may be involved in the organization of subsequent parts of the movement (see the Conclusion). It should be noted, however, that Slaughter et al. (1970) found no changes in the pattern of the periodic discharge activity of cerebellar dentate (and cortical) neurons in three patients suffering from infantile athetosis during the performance of spontaneous active or passive movements (light anesthetic).

Of special importance is the study of the mechanisms of the generation of cerebellar discharge activity associated with voluntary movement. It is known that the activity of cerebellar output nuclei is controlled by the action of the Purkinje cells, which in turn are controlled by climbing and mossy (through the granule cells) fibers. In a long series of studies, it was shown that the Purkinje cells generate two types of spikes: "simple" spikes, which are produced by the mossy fibers -- granule cell system -- and "complex" spikes, which are produced by impulses in the climbing fibers issuing from the inferior olivary neurons (for a survey, see Thach, 1972). While recording the activity of the Purkinje cells in the intermediate zone of the cerebellar hemisphere in monkeys, Thach utilized this fact in forming a theory on the activity of the

Purkinje cells and their afferent inputs--the climbing and mossy fibers. Since each Purkinje cell receives one climbing fiber originating from a neuron in the inferior olives, the presence of a complex spike served as an indicator of the activity of the inferior olivary neurons. Thach (1968) determined that in a quietly seated monkey, complex spikes in Purkinje cells are observed at irregular intervals and at a low frequency. During rapidly alternating movements, no movement-related changes were detected in the pattern of these complex spikes, even in cases in which the simple spikes of the same Purkinje cell were clearly movement-related. However, during a rapid conditioned reflex movement in response to a light flash (Thach, 1970b), a complex spike was observed during or shortly prior to movement in 30% of those (and only those) Purkinje cells that exhibited a movement-related simple spike. A complex spike (and thus the discharge of an inferior olivary neuron) was observed during movement in one direction, but not during movement in the opposite direction. This spike could appear prior to or during the movement. The presence of a complex spike has no effect on the appearance of a simple spike.

Purkinje-cell discharges in the intermediate zone of the anterior portion of the cerebellum during postural fixation or the performance of rapid movements has features in common with the discharges of neurons of the intermediate nucleus to which the Purkinje cells project. Both types of cells have a high-frequency tonic discharge, and during movement, both the Purkinje cells and the cells of the intermediate nucleus modify their discharge at the same instant and in the same direction (frequency more often increased than decreased). Since the Purkinje cells have an inhibitory effect on cerebellar nuclear neurons, Thach surmises that Purkinje cells very probably have an inhibitory influence on cerebellar nuclear activity already evoked in connection with a voluntary movement. On the other hand, an analysis of his results suggests that the impulse frequency in mossy and climbing fibers is modified prior to movement, thereby modifying the activity of the Purkinje cells. In turn, the frequency variation in these fibers may be the result of descending effects from the precentral motor cortex.

It is therefore very likely that changes in the activity of Purkinje cells prior to movement are the result of effects from the contralateral precentral motor cortex, which is indirectly connected with the Purkinje cells by the mossy and climbing fibers. Since the climbing fibers originate from the inferior olivary neurons, it is entirely possible that the activity of these neurons is also modified by a voluntary movement (prior to or during movement).

Summarizing his data on neuronal activity, cerebellar connections, and the behavioral defects produced by removal of the cerebellum, Thach draws two conclusions: (1) the cerebellum aids in the initiation and performance of some types of movement and postural fixation; (2) the cerebellar output cells with a high discharge frequency are apparently specifically related to motor performance and postural fixation. In addition, Thach notes that with the exception of differences in sustained frequency in the absence of movement, the behavior of cerebellar output neurons is similar to that described under analogous conditions of postural fixation and voluntary motor performance for neurons of the precentral cortex (Evarts, 1966, 1969), the ventrolateral thalamic nucleus (Evarts, 1970), the pallidum (De Long, 1971, 1972), and even for spinal motoneurons. This finding suggests that all these components of the motor system may have an aggregate (as opposed to individual or accessory) function in the control of movement. There remains, however, the question: what specific additions and what unique contributions are made by each component to the organization of voluntary movement?

If cerebellar output plays a role in initiating the activity of pyramidal tract neurons prior to movement, we must then determine which portions of the cerebellum are involved in initiating this activity and how these structures are activated. Since the cerebellum can affect the motor cortex through the ventrolateral thalamic complex, it is quite likely that the activity of thalamic neurons prior to movement (Jasper and Bertrand, 1966; Evarts, 1970) is initiated by earlier activity of the cerebellum. These projections from the cerebellum to the motor cortex originate essentially from the lateral portions of the cerebellar cortex, which are themselves excited by the cerebral cortex, particularly its "association" zones. These "association" regions of the cortex can thus affect (primarily through the cerebropontine pathway) the lateral portions of the cerebellum, which in turn can activate the PTNs in the cerebral motor cortex. Hence, owing to their connections with the cerebral motor cortex, the lateral portions of the cerebellar hemispheres and the dentate nucleus may participate in the initiation and regulation of voluntary movements according to the mechanism predicted by Holmes (1930).

However, data obtained by V. B. Brooks et al. (1972) do not appear to be consistent with such a hypothesis. These authors found that during brief cooling of the dentate nucleus, alternating conditioned movements in monkeys are performed with greater speed than under control conditions. However, an increase in the maximum speed of movement is not accompanied by a rise in the phasic discharge frequency of the cortical neurons. Consequently, control by the dentate nucleus in this situation is not necessarily related to its primary action on neurons of the precentral cortex.

On the other hand, it is known that the intermediate portions of the cerebellum are also connected to the ventrolateral thalamus and the reticular and vestibular nuclei, but principally to the red nucleus (especially its magnocellular portion). Cerebellar projections overlap with endings from the cerebral cortex (or primarily from the primary motor cortex in monkeys: see Evarts and Thach, 1969) in that region of the red nucleus from which descending pathways emanate into the spinal cord. One of the inputs to the intermediate (spinal) portion of the cerebellum consists of a projection from the cerebral cortex, but unlike the lateral portions of the cerebellum, it consists principally of fibers from the somatomotor region rather than the "association" region. This means that the intermediate portions of the cerebellum may be initially activated by input from the motor cortex (through the large corticopontine and corticoreticular pathway), while the cerebellum can act on the spinal cord through the red nucleus. Another pathway emanates from somatosensory receptors up through the spinal cord to the intermediate portion of the cerebellar cortex. This projection may serve to provide feedback, informing the cerebellum of the effects of its actions on the spinal cord via the red nucleus. Thus, the intermediate portion of the cerebellum can act on the red nucleus while being acted on by the somatomotor cortex, thereby participating in the initiation and regulation of movement. This corresponds to the mechanism postulated by Herrick (1924).

Systematic studies have not yet begun on the spinal neuronal mechanisms of the organization and regulation of voluntary movement in animals. The following chapters will describe the results of our own studies on the spinal mechanisms of the organization of voluntary movement in man. These studies have shown that the initiation of a voluntary movement is preceded by complex and prolonged changes in the state of various spinal interneuronal systems and in the motoneuronal nuclei of future agonists and antagonists.

CHAPTER 2

NATURE OF THE H-REFLEX AND METHODS FOR ITS ELICITATION

The H-reflex method is the basic method employed in our studies of the spinal mechanisms of the organization of voluntary movement. This chapter will therefore begin with a discussion of the experimental data obtained through the use of this method.

Applying a weak electrical stimulus to the tibial nerve (TN) in the popliteal fossa of human subjects, P. Hoffmann (1918, 1922) recorded a synchronized electrical response of the gastrocnemius muscle (GM)* with a latency of 25-30 msec. When the strength of the electrical stimulus was increased, the amplitude of this response also increased, and a second response appeared with a latency of only 5 msec. A further increase in the strength of the electrical stimulus led to a rise in the amplitude of the short-latency (early) response and a decrease in the amplitude of the late response. When the stimulating electrode was more proximally positioned, the latency of the early response increased, while that of the late response decreased. The amplitude of the late response increased during voluntary GM contraction and decreased during voluntary agonist contraction as well as after conditioning stimulation of the TN. In each of these cases, there was no change in the amplitude of the early GM response.

On the basis of his data, Hoffmann concluded that the early GM response is associated with the stimulation of efferent motor nerve fibers. This response was termed the "M" ("motor") response. Hoffmann associated the late GM response with the reflex response of the motoneurons of this muscle to electrical stimulation of the afferent TN fibers. This reflex was termed the "H-reflex" (in honor of Hoffmann), while the electrical manifestation of this reflex in the GM was called the "H-response" (Magladery and McDougal, 1950). Struck by the similarity of the pattern and latency of a

*This abbreviation will be used whenever the recording electrodes are positioned so as to record the electrical activity of all three heads of m. triceps surae.

reflex gastrocnemius H-response evoked by electrical stimulation of the popliteal nerve to that of the response of this muscle to a blow to the Achilles tendon (the Achilles tendon reflex), Hoffmann surmised that both responses are a manifestation of the same stretch reflex and may be counted among the simplest monosynaptic reflexes.

Direct evidence of the existence of spinal reflexes with a central synaptic relay was later provided by Eccles and Renshaw (see Kostyuk, 1959; Eccles, J., 1957). Lloyd (1943a) showed that electrical stimulation of a muscle nerve with a voltage sufficient to evoke the excitation of only low-threshold, fast-conducting Group Ia afferents is sufficient for monosynaptic activation of the motoneurons.

Direct evidence of the monosynaptic nature of the electrically evoked GM H-reflex in man was provided in studies by Magladery et al. (Magladery et al., 1951a; Magladery, 1955). Inserting several needle electrodes at interspinous intervals into the vertebral canal of volunteers under superficial barbiturate anesthesia, they recorded potentials from the spinal cord during evocation of the H-reflex. Magladery and his associates stimulated the TN and recorded at various levels of the lumbrosacral section of the spinal cord two synchronized potentials: the R- and A-potentials. According to the conclusion reached by the authors, the R-potential is associated with the propagation of an afferent posterior root volley, the A-potential with reflex motoneuronal discharge. The following facts tend to support the conclusion regarding the afferent nature of the R-potential: (1) the latency of the R-potential increases when this potential is recorded in more rostral sections of the spinal cord; (2) the R-potential climbs when the electrical stimulus is strengthened; (3) the R-potential is abolished after 20 min of ischemia (by tourniquet) of the limb, along with the H-response, while the M-response is not appreciably reduced (see Section 2). Magladery and his associates cite the following data as evidence of the efferent nature of the A-potential: (1) the latency of the A-potential decreases when this potential is recorded in more rostral sections of the spinal cord; (2) on evocation of a second H-reflex a short time after the first (according to the records presented by these authors, about 12-50 msec), there is a sharp depression of the second H-response and the A-potential is abolished (see Chapter 3, Section 4); (3) when nerve stimulation is strengthened, there is a decline in both the H-response and the A-potential (see Section 1). In this case, the authors attribute the decline in A-potential to the encounter of the antidromic volley in the efferent fibers with the centrifugal reflex volley of the somewhat more proximally positioned recording electrode. They attribute the fact that the antidromic volley itself was not recorded to its temporal dispersion due to differences in the conduction velocity of different motor fibers.

The basic result of Magladery's study was the discovery of a central delay for the H-reflex. At level L_4, approximately 1.5 msec pass between the onset of the posterior-root afferent volley (R-potential) and the appearance of the reflex motoneuronal discharge (A-potential). If we take into consideration the conduction time along the proximal areas of the posterior and anterior roots up to the recording level and the duration of intraspinal conduction, sufficient time is left for only one synaptic delay.

During the past 10 to 15 years, Hoffmann's method has gained wide acceptance in clinical physiological research (for a survey, see Baikushev et al., 1974). Yet there remain a great many unanswered procedural questions and obscure details concerning the nature of the H-reflex. This chapter will deal with these unresolved aspects of the H-reflex method.

1. DEPENDENCE OF THE AMPLITUDE OF THE H-RESPONSE ON THE STRENGTH OF NERVE STIMULATION

When separate recordings are made of the H-response of the soleus muscle (Sol) and of the medial gastrocnemius muscle (MG), some pronounced differences are revealed. In this section, only the H-responses of the Sol will be described. The special features of the H-responses of the MG will be examined in detail in Section 4.

For clarity, the curve in Fig. 1E relating the amplitude of the Sol H-response to the strength of TN stimulation has been divided into three zones. In zone I, a slight increase in the strength of the stimulus above threshold produces a substantial increase in the amplitude of the H-response. Stimuli that elicit H-responses in zone I will henceforth be referred to as "suprathreshold stimuli." Suprathreshold stimuli evoke either no M-response of the Sol or a very slight M-response many times smaller than the H-response (Figs. 1A-D). In the latter case, a gradual increase in the strength of the suprathreshold stimulus produces only a very slight increase in the amplitude of the M-response.

Zone II includes the submaximal and maximal (in terms of amplitude) H-responses of the Sol. Within zone II, an increase in the strength of the stimulus has a relatively slight effect on the amplitude of the Sol H-response. The maximal stimuli for the H-response either elicit a threshold M-response or increase an M-response that may have already emerged in zone I. In both cases, a gradual increase in the strength of the stimulus in zone II produces a relatively small increase in the M-response (less than in zone III, but more than in zone I).

In zone III, a gradual increase in stimulus strength leads initially to a rapid decrease in the amplitude of the H-response,

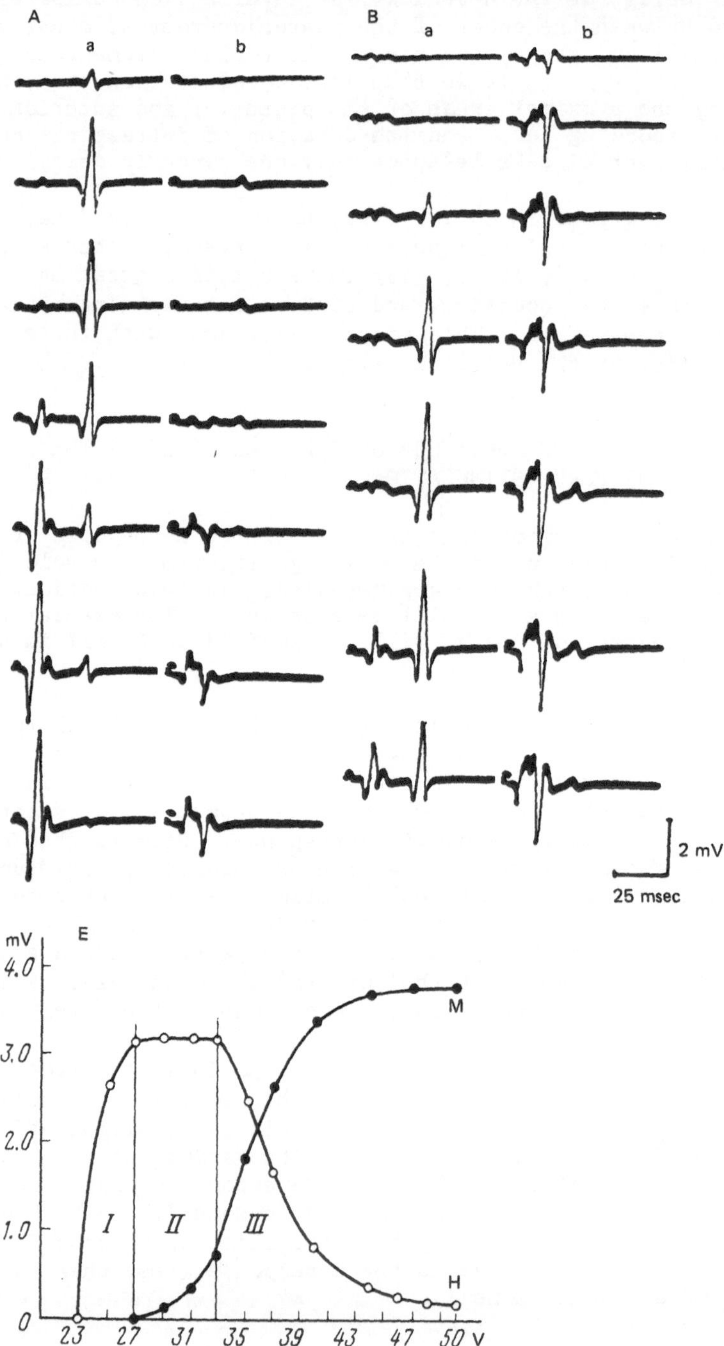
A
a
b
B
a
b
2 mV
25 msec
mV
E
4.0
3.0
2.0
1.0
0
I
II
III
M
H
23
27
31
35
39
43
47
50
V

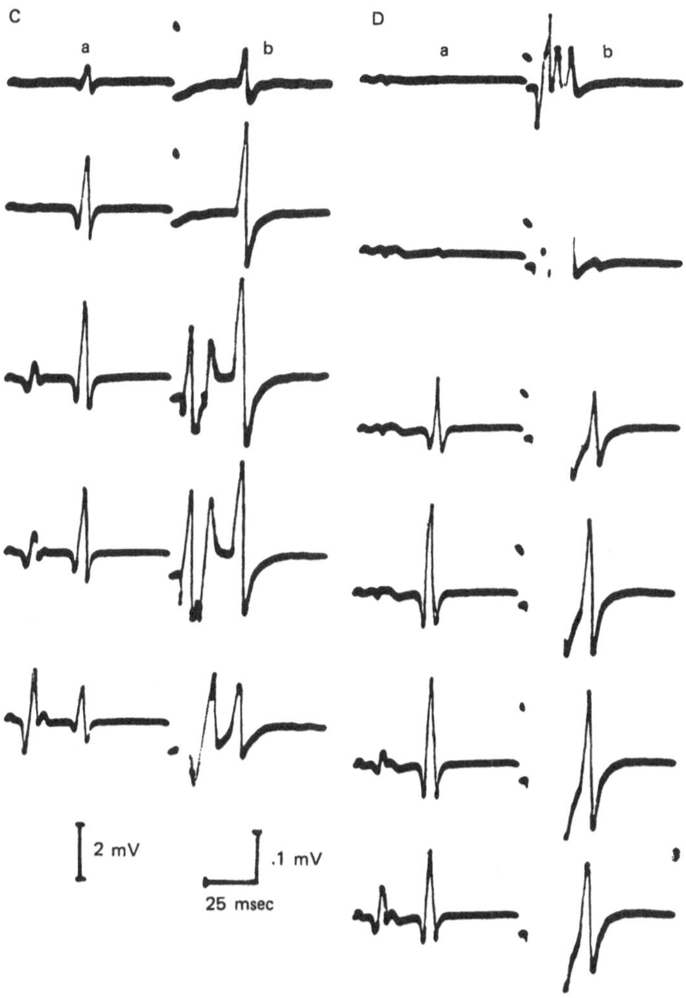

Fig. 1. H- and M-responses of the Sol (a) and the ATM (b) for four different positions of the stimulating electrode. (A) Electrode positioned more medially than B. (C) Electrode positioned more medially than D. Strength of nerve stimulus increases from top to bottom. Calibration below B applies to records A and B. Calibration below C applies to records C and D. (E) Plots of amplitudes of H- and M-responses (in mV) vs. strength of stimulus (in V). Explanation in text.

followed by a slow reduction in the rate of decrease; this is accompanied by a parallel increase in the amplitude of the M-response. The curve relating the amplitude of the H-response to stimulus strength in zone III is practically a mirror image of the curve of the M-response. The maximal M-response is evoked by a stimulus strength approximately twice that required to elicit the threshold M-response.

A convenient method of relating the two responses is to represent the control amplitude of the H-response as a percentage of the amplitude of the maximal M-response, i.e., the ratio H/M_{max}. Such a representation of the magnitude of the H-response has a real physiological significance, since it permits certain judgments to be made concerning the ratio of excited motor units to their total number in a given muscle.

The ratio of H- to M-responses and the dependence of H- and M-responses of the Sol on the strength of the stimulus to the TN are significantly influenced by the position of the stimulating electrode in the popliteal fossa (see Fig. 1). The experimental results given in Fig. 2 indicate that the steepness of the slope of the rising portion of the curve "strength of stimulus vs. amplitude of H-response of Sol" in zone I may vary, depending on the position of the stimulating electrode. The slope is generally less steep for more medial stimulating electrode placements. The shape of the curve "strength of stimulus vs. amplitude of H-response of Sol" in zone II and the width of this zone also vary for different positions of the stimulating electrode. The presence of a distinct plateau in zone II is more characteristic of a proximo-lateral placement in the popliteal fossa.

In the same experiment, the maximal amplitudes of the H- and M-responses (and thus the ratio H_{max}/M_{max}) may also be different, depending on the position of the stimulating electrode (see Figs. 1A-D). These differences can be understood if we consider that the amplitude of the H-response does not depend only on the number of electrically excited afferent (for the H-response) fibers. If an increase in the number of excited H-reflex afferents within certain limits leads to an increase in the H-response, then an increase in the number of simultaneously excited efferents leads to a depression of the H-response (see below). It must be borne in mind that when the active electrode is applied to the popliteal fossa, stimulation of the sciatic nerve and two of its branches--the TN and the peroneal nerve (PN)--is possible. It is quite likely that the ratio of afferent to efferent H-reflex fibers and their proximity to the surface are different in the tibial and sciatic nerves. Moreover, there is the possibility that afferent fibers both from the foot extensors and from other muscles comprising the PN--particularly the anterior tibial muscle (ATM)--will be involved in the synaptic activation of the motoneurons of these

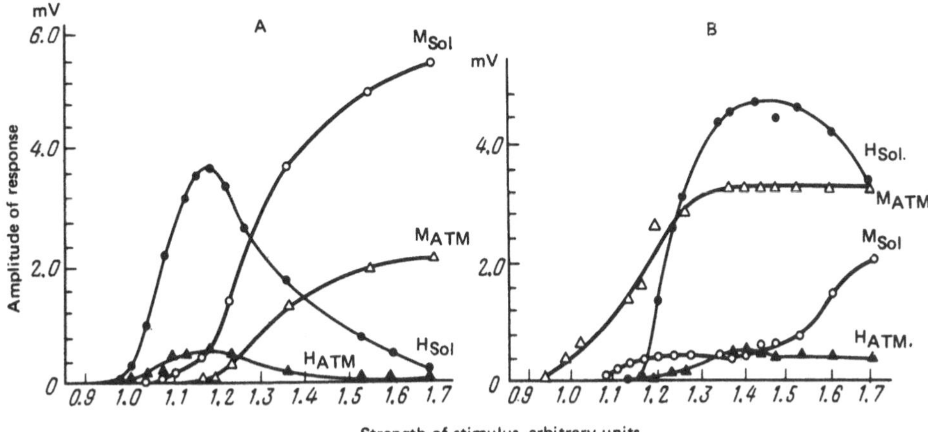

Fig. 2. Amplitudes of the H- and M-responses of the Sol as a function of the strength of nerve stimulation for two different positions of the stimulating electrode. (A) Electrode positioned more medially than B.

extensors (see Chapter 4, Section 2). These factors combine to make the amplitudes of the H- and M-responses dependent on the position of the stimulating electrode.

As mentioned earlier, the increase in the strength of the nerve stimulus that leads to a considerable rise in the amplitude of the M-response is accompanied by a decrease in the amplitude of the H-response (zone III). P. Hoffmann (1922) suggested that the decrease in the amplitude of the H-response in the presence of a strong M-response is due to the fact that the reflex motoneuronal volley is blocked by an encounter with the antidromic volley in the motor fibers (or in the bodies of the motoneurons). This view has been shared by other authors (Magladery et al., 1950).

However, Hoffmann's hypothesis regarding an antidromic blockade as the only cause of the depressed H-response during "supramaximal" stimulation contradicts certain facts that indicate that antidromic blockade does not occur in all motor axons (or motoneuronal bodies). First, even during maximal M-response (the activation of all motor fibers) during rest, complete depression of the H-response does not occur (see Fig. 1A). Second, under conditions of voluntary facilitation, it is possible to obtain a strong H-response that is preceded by the maximal M-response (Fig. 3A, traces 1 and 2). Third, under conditions of voluntary facilitation, an H-response can be obtained with an amplitude

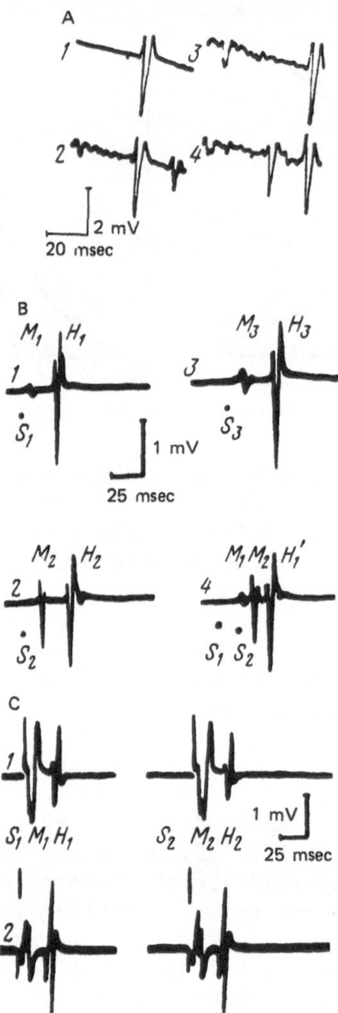

Fig. 3. H- and M-responses of the Sol during voluntary foot extension, and to two stimuli during rest. (A) 1 - max. M-response during rest; 2 - max. M- and supramax. H-response during movement; 3 - max. H-response (= max. M-response) during movement; 4 - M-response (∿50% of max.) and max. H-response during movement; (B) H-responses to single stimulus (1-3) and two (4) stimuli. Strength of stimulus, V: 25.5 (1), 31 (2), 27 (3). (C) M- and H-responses of two sections of MG (1,2) to S_1 (22 V, left) and S_2 (22.5 V, right).

approximately equal to the maximal M-response, even if this H-response is preceded by a strong M-response (Fig. 3A, traces 3 and 4). To account for these facts, it must first be assumed that some of the motor axons have a relatively high conduction velocity that permits the anti- and orthodromic (reflex) volleys in them to diverge (Lloyd, 1943b).

In our experiments, two procedures were used to demonstrate that a depression of the H-response during supramaximal stimulation (in zone III) cannot be explained solely by an antidromic blockade (first hypothesis), but must rather be due to the actions of a central inhibitory mechanism (second hypothesis).

<u>First procedure</u> (Fig. 3B): A stimulus S_1 was chosen that evoked a submaximal H-response H_1 (trace 1), and a supramaximal stimulus S_2 that evoked a stronger M-response and H-response H_2 (trace 2) with an amplitude somewhat less than the maximal H-response H_3 (trace 3). In the experiment with two stimuli (trace 4), stimulus S_2 was applied 11 msec after S_1. An interval of 11 msec was chosen in order to (1) ensure an encounter between the antidromic volley in the motor fibers from S_2 and the orthodromic reflex volley evoked by S_1, and (2) prevent the interference of the M_2- and H_2-responses with the H_1-response. The modified H_1-response produced by the use of the two stimuli is designated in the figure as H_1' (trace 4).

The logic behind this procedure is as follows: If, on paired stimulation, H_1' is stronger than H_2, the first hypothesis would be proved false. Indeed, if the fact that H_2 is weaker than H_1 were due solely to antidromic blockade (first hypothesis), H_1' then should be weaker than or as strong as H_2 in the experiment with paired stimuli, since: (1) in the two-stimulus experiment, an antidromic blockade is deliberately produced that is no less than that in the S_2 experiment, and (2) the S_1 stimulus excites no more efferent fibers than stimulus S_2 (afferent stimulation by S_2 cannot affect H_1' in the case of two stimuli). Nevertheless, the experiments showed that H_1' can be stronger during paired stimulation than H_2 during stimulation by S_2 only.

Hence, the decrease in H_2 is due not only to antidromic blockade, but to other factors as well. If we assume that central inhibition is involved in the phenomenon observed, then the fact that H_1 is stronger than H_2 can be accounted for in the following way: During application of two stimuli, the excitation induced by the S_2 stimulus fails to reach the spinal cord and there is a blocking of the reflex discharge evoked by the S_1 stimulus only in antidromically excited axons. During stimulation by S_2 only, the volley evoked by the stimulus apparently has time to reach the spinal cord (e.g., via the faster-conducting motor axons or I-b afferents) and excite (presumably) the inhibitory Renshaw cells or the inhibitory interneurons of Cajal's nucleus, which mediate the effects of the I-b afferents on the motoneurons. Thus, during facilitation of the motoneurons by primary afferent volley, these inhibitory interneurons can inhibit the discharge of those motoneurons the axons of which were not antidromically excited.

Second procedure: This is based on the assumption that if central inhibition is at all involved in the depression of the H-response, a decrease in the H-response can be observed during a constant M-response; this would suggest central inhibition of the reflex response of those motoneurons the axons of which were not antidromically excited. The records of one of the experiments shown in Fig. 3C point to such a possibility. Two pairs of recording electrodes were applied to two different portions of the MG muscle (1 and 2). Supramaximal stimulation of the nerve by S_1 elicited M_1- and H_1-responses (left traces). Then the strength of the stimulus was very slightly increased (S_2), such that the M-response did not change below one of the recording electrode pairs (M_2, upper right trace). Inspection of a number of records revealed a decline in the H_2-response despite no change in the M_2-response. At the same time, there is an increase in the M-response on the adjacent muscle section in response to the stimulus increase from S_1 to S_2 (cf. left and right lower traces).

The presence of fast-conducting motor fibers in the stimulated nerve makes it possible for the inhibitory Renshaw cells to participate in the mechanism of the depression of the supramaximal H-response as they are antidromically activated through these motor fibers. The prerequisite for the participation of the Renshaw cell system in this depression of the H-response is that the conduction velocity in the fastest efferent fibers be somewhat higher than in the slowest afferent fibers participating in reflex motoneuronal activation. This appears to be quite probable in the light of experimental data (Eccles, J. C., et al., 1961). In addition, central inhibition of the H-reflex may possibly be accomplished by a system of inhibitory interneurons of Cajal's nucleus, which are excited by an orthodromic volley in the fast I-b afferent fibers from the tendon organs of Golgi (Eccles, J., 1959).

The participation of the inhibitory interneuronal system of Cajal's nucleus in the inhibition of the H-reflex during supramaximal stimulation requires that the conduction velocity in the fastest I-b afferents be slightly higher than in the slowest I-a afferents. This is very likely the case. According to data obtained by Lundberg and Winsbury (1960), there is no difference in the conduction velocity of I-a and I-b afferents in the nerves of the ankle muscles in the cat, but according to data obtained by Jansen and Rudjord (1964), the conduction velocity in afferents from the soleus tendon organs is as high as 116 m/sec, i.e., the speed of the fastest I-a afferents.

The data obtained thus show that when the H-response is evoked in zone II or III and preceded by an M-response, the magnitude of the reflex H-response can be reduced not only by antidromic blockade, but also by central intraspinal inhibition. Hence, the degree of depression of the amplitude of the H-response during supramaximal stimulation can depend on the state of the spinal inhibitory system (the Renshaw cells, perhaps, or the inhibitory interneurons of Cajal's nucleus), and the supramaximal H-reflex of zone III can be used to evaluate the functional state of this system (see Chapter 3, Section 3).

2. AFFERENTS OF THE H-REFLEX

Considering data on a direct relationship between the sensitivity of nerve fibers to ischemia and fiber diameter (Gasser and Erlanger, 1937; Fox and Kenmore, 1967), Magladery et al. (1950) compared changes in the reflex H-response and the peripheral M-response during tourniquet-induced ischemia of the lower limb. Since these authors observed that ischemic depression of the H-response occurred earlier than that of the M-response, they concluded that during the early stages of ischemia, conduction along afferent H-reflex fibers is preferentially blocked while conduction along motor fibers remains relatively intact, and therefore that afferent H-reflex fibers are thicker than motor fibers. This conclusion is consistent with the fact described by P. Hoffmann (1918) that a strong H-response can be evoked without the simultaneous excitation of motor fibers (without an M-response).

However, the methodological and conceptual flaws in Magladery's experiments required that they be repeated and broadened (Kots, 1968, 1970). First, in Magladery's experiments, the tourniquet cuff was positioned more proximally than the stimulating electrode (proximal tourniquet). In this case, the amplitude of the M-response reflects only the state of excitation and conduction of motor fibers at a point distal to the tourniquet. To utilize the M-response to determine the participation of an efferent blockade in the depression of the H-response, the speed of formation of a blockade in the efferent motor fibers at the point of tourniquet pressure would have to be compared with the speed at a more distal point. This was achieved by comparing changes in the M-response for a tourniquet first positioned proximally to the stimulating electrode (proximal tourniquet) and then distally to the stimulating electrode (distal tourniquet). Second, there is a possibility that the depression of the H-response during ischemia is due not only to a blockade of afferent fibers, as Magladery and his associates assumed, but also to ischemic blockade of the muscle spindles (Matthews, B. H. C., 1933; Paintal, 1959; Eldred et al., 1960), which can lead to a reduction of the excitability of the motoneurons and thus to a depression of the H-reflex.

To determine how ischemic deafferentation affects the level of reflex excitability of deafferented motoneurons, we employed a distally applied tourniquet to ensure deafferentation due to blockade in the afferent fibers (and possibly the blockade of muscle receptors) at a point distal to the tourniquet. At the same time, stimulating electrodes positioned at an intact nerve portion proximal to the tourniquet provided a constant test stimulus throughout the ischemia, thereby allowing an evaluation of the state of excitability of the deafferented motoneurons.

As our experiments show, there is a nearly uniform decrease in the amplitude of the M-response for the first 25-30 min after application of a proximal (Fig. 4A) and a distal (Fig. 4B) tourniquet. The weak M-response produced by the excitation of the lowest-threshold motor fibers begins to diminish after 5-7 min of ischemia for both proximal and distal tourniquet positions (Fig. 5B). The maximal M-response begins to diminish during about the 10th minute of ischemia, usually falling to approximately 95% of its original amplitude after 12-15 min (see Figs. 4A,B). Over the next 10 min, the maximal M-response continues to diminish gradually, such that its amplitude is 80-85% of the original value after 23-25 min of ischemia. At this point, the amplitude of the M-response begins to decrease rapidly, falling somewhat faster during distal than proximal ischemia (Fig. 6A). After 40-42 min, the M-response almost completely disappears.

Thus, during the first phase of ischemia (∿25 min), an efferent blockade develops at a more or less identical speed at and distal to the point of tourniquet pressure, which permits a judgment to be made regarding the participation of efferent blockade in the depression of the H-response based on the decrease in the M-response during proximal ischemia. Moreover, the pressure in the region of the tourniquet creates several additional conditions for a deepening of the blockade, a factor that evidently helps account for the somewhat faster decrease in the M-response during the late phase (25-40 min) of distal ischemia than during proximal ischemia.

As mentioned earlier, the effect of tourniquet ischemia on the reflex excitability of motoneurons can be determined during distal ischemia. During the first 3-5 min following distal application of the tourniquet, the suprathreshold control H-response exhibits no noticeable changes. Thereafter, it begins to increase, gradually at first, then more rapidly (see Fig. 4C). If the intensity of the test stimulus is reduced after the increase in the amplitude of the H-response (in this case, the 16th minute of ischemia), the new, weaker H-response will again begin to increase (see Fig. 4C; at 21 min, it was necessary to again reduce the strength of the stimulus). Using as the control an H-response with an initial amplitude approximately 50% of the maximal and recording

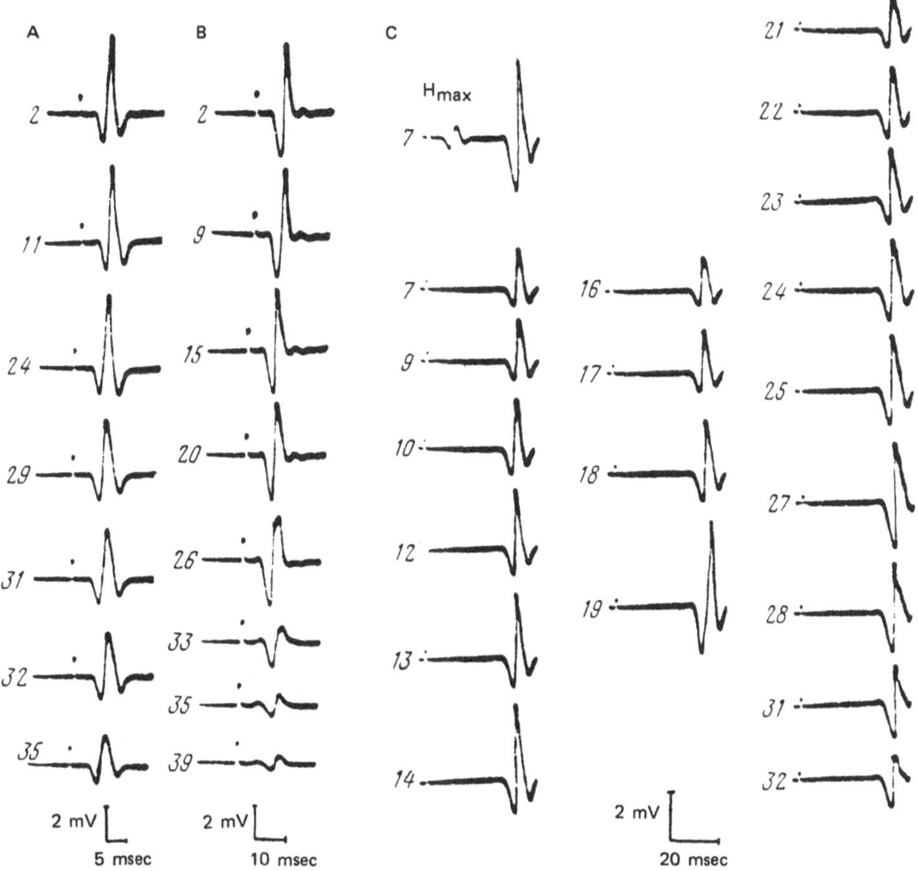

Fig. 4. Maximal M-response of Sol during proximal (A) and distal (B) ischemia and suprathreshold H-response of Sol during distal ischemia (C). Numbers beside records indicate time elapsed since application of tourniquet in minutes. The upper record under C gives the maximal H-response at the 7th minute of the ischemia.

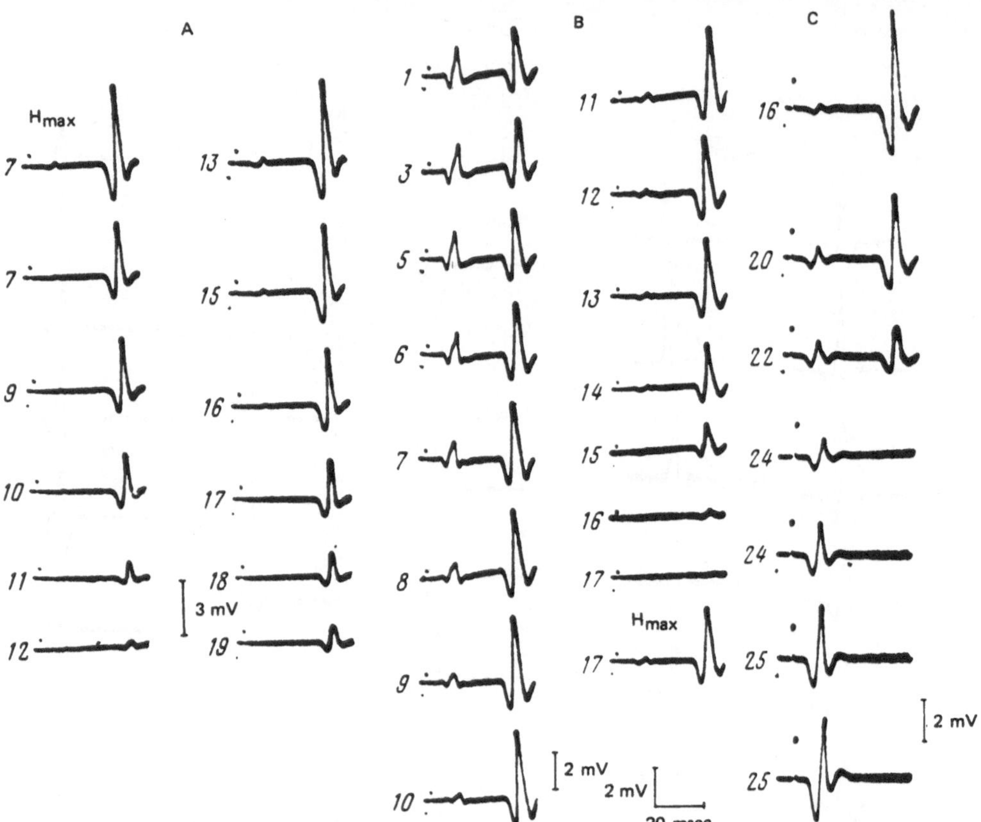

Fig. 5. Suprathreshold (A), supramaximal (B), and maximal (C) H-responses during period of proximal tourniquet ischemia. Numbers beside records indicate duration of ischemia in minutes. The upper record under A gives the maximal H-response at the 7th minute of the ischemia period; the lower record under B gives the maximal H-response at the 17th minute of ischemia elicited by an increase in stimulus strength; C indicates the maximal H-response elicited at the 16th, 20th, and 22nd minutes of ischemia during incremental increases in stimulus strength. Subsequent increases (24th, 25th minute) elicited no H-responses.

the time at which the maximal amplitude was achieved, we determined that the amplitude of the H-response increases most rapidly during the period extending from the 8-10th minute to the 20-23rd minute of ischemia. After this period, there is a decrease in the rate of growth. From the 25th to the 27th minute, the amplitude of the H-response decreases in direct proportion to the decrease in the amplitude of the maximal M-response. If we choose as control the maximal or supramaximal (zone III) H-response, an increase in its amplitude is observed 8-10 min after application of the distal tourniquet. At the 15-17th minute, the amplitude of the H-response may somewhat exceed the maximal amplitude of the H-response at the start of ischemia. In the interval from 18-20 to 23-25 min, the amplitude of such an H-response does not noticeably change. After the 23-25th minute, the amplitude begins to decrease, the decay paralleling that of the M-response.

Thus, far from producing a depression of the reflex excitability of deafferented GM motoneurons, ischemic deafferentation actually leads to a significant increase in their reflex excitability that accelerates with the duration of the ischemia period.

The situation is reversed during proximal tourniquet ischemia (the technique used by Magladery et al.). During the first 5-7 min, there is no appreciable change in the control H-response, which, however, begins to decline--slowly at first, but quite rapidly after the 8-10th minute (see Fig. 5A). The smaller the initial amplitude of the control H-response, the earlier and more complete the depression of the H-response. If the strength of the test stimulus is increased during the period when the suprathreshold control H-response is already sharply reduced (12th minute of ischemic period), it is possible to again elicit a strong H-response that may even approach the maximal value (13th minute of ischemic period). The amplitude of the new control H-response immediately begins to fall; the smaller the amplitude, the greater the rate of fall.

The change in the maximal or supramaximal H-response is somewhat different. From the 5th to the 7th minute of a proximal ischemia, there is a slight increase in the amplitude of the maximal H-response and a more noticeable increase in the amplitude of the supramaximal H-response, accompanied by a slight, parallel decrease in the amplitude of the preceding weak M-response (see Fig. 5B). At the 9-10th minute, the amplitude of the supramaximal H-response may even exceed somewhat the initial maximal amplitude of the H-response. One explanation for this increase is that ischemic occlusion of the motor fibers (indicated by the decrease in amplitude of M-response) leads to a depression of the antidromic volley, whereby the inhibitory action of this volley on the reflex response of the motoneuronal nucleus is weakened (see the preceding section). Following the period of growth, the amplitude of the

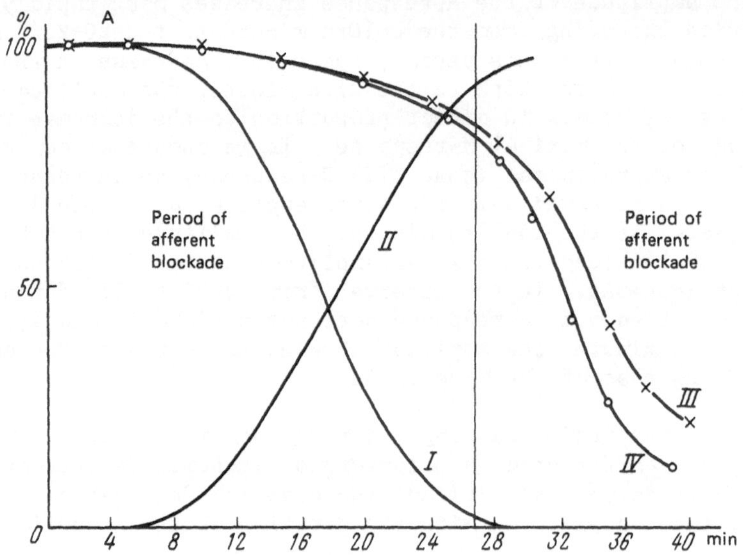
A
%
100
50
0
Period of afferent blockade
Period of efferent blockade
I
II
III
IV
4
8
12
16
20
24
28
32
36
40
min

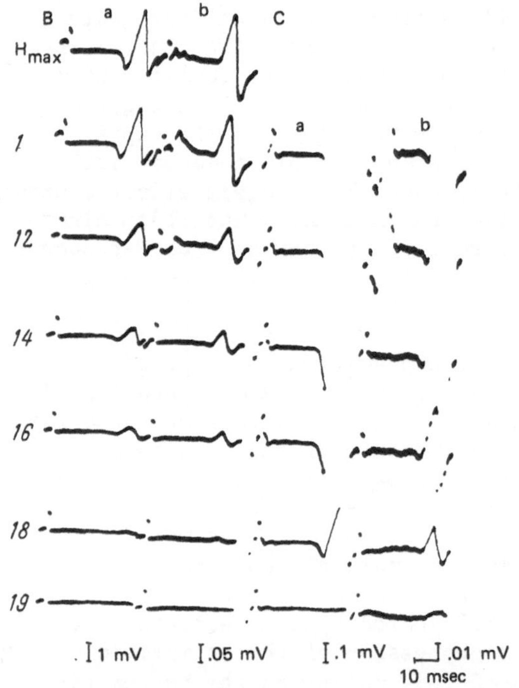
B
a
b
C
D
E
H_{max}
1
12
14
16
18
19
25
1 mV
.05 mV
.1 mV
.01 mV
10 msec
1 msec

control H-response begins to fall rapidly and is sharply reduced by the 16-20th minute. However, if stimulus strength is increased at the 16-18th minute of the ischemic period, it is possible to again elicit a strong H-response with an amplitude up to 80% of that of the maximal H-response at the start of ischemia (see Fig. 5B, last record). If the ischemia exceeds 22-25 min, no H-response can be elicited, regardless of the strength of the stimulus (Fig. 5C).

A proximal ischemia thus leads to a significantly faster decrease in the H-response than in the M-response. The supra-threshold H-response almost disappears when the maximal M-response is still approximately 95% of the initial response. During the period of the fastest decline of the amplitude of the H-response, culminating in its extinction, the maximal M-response diminishes by only 15-20%. Judging by the rate of the decline of the H-response, the fastest increase in the number of blocked H-reflex afferents appears to occur from about the 10th to the 25th minute of the ischemic period. Judging by the decline of the maximal M-response, the fastest increase in the number of blocked efferent motor fibers begins after the 20-25th minute of ischemia. The first 25 min of the ischemic period could thus be designated the period of preferential afferent blockade, and the period thereafter as the period of efferent blockade. This is shown graphically in Fig. 6A.

That the H-response declined more rapidly than the M-response may be explained by the greater sensitivity of the bulk of the afferent H-reflex fibers of the GM to ischemia than the motor fibers of the GM. Other possible explanations for the decline of the H-response--such as (1) a blockade of the reflex volley in the

---

Fig. 6. Effects of ischemia. (A) Reconstructed curves showing development of afferent blockade (I) and increase in reflex excitability of spinal motoneurons (II), and curves showing development of efferent blockade (III, proximal; IV, distal tourniquet) during tourniquet ischemia of lower limb. Abscissa: duration of tourniquet. Ordinate (curves III and IV): amplitude of maximal M-response in percentage of maximal M-response at start of ischemic period. (B,C) Submaximal H-response of Sol (a) and ATM (b) during proximal ischemia. The records under C were made with greater amplification than those under B. Upper record under B indicates maximal H-response at start of ischemia. (D,E) Maximal M-response of MG at different stages of proximal ischemia. The records under E were made at a higher amplification and higher sweep speed than those under D. Numbers beside records indicate duration of tourniquet in minutes.

motor axons at the point of tourniquet pressure not revealed by the M-response during proximal ischemization, or (2) a reduction of the reflex excitability of ischemically deafferented motoneurons--were rejected on the basis of the results of the experiments described earlier. Taking into account experimental data on the relationship between the sensitivity of nerve fibers to ischemia and fiber diameter, it can be concluded that a significant portion of the afferent fibers in the H-reflex arc of the GM have a greater diameter than most of the motor fibers. Such a conclusion is in agreement with other facts: (1) the H-response usually has a lower threshold than the M-response; (2) the conduction velocity along the afferent limb of the H-reflex is greater than along the efferent limb (Magladery and McDougal, 1950; Magladery et al., 1951; Diamantopoulos and Gassel, 1965).

Moreover, the results of experiments with tourniquet ischemia have pointed to the heterogeneity of the H-reflex afferent limb. Indeed, the greater the strength of the afferent test stimulus, the greater the ischemic resistance of the H-response elicited by the stimulus; and after the control H-response has disappeared, it can be reelicited for a time by increasing the strength of the stimulus (see Fig. 5). One explanation for these facts might be that ischemia leads to a gradual increase in the threshold of the H-reflex afferents to electrical stimulation. But experimental data contradict such an interpretation (Maruhashi and Wright, 1967). The direct dependence of the ischemic resistance of the H-response on the strength of the test stimulus is apparently due to the fact that an increase in stimulus strength is accompanied by the recruitment of higher-threshold (thinner, and thus more resistant to ischemia) afferent fibers that, along with the relatively low-threshold afferents, are incorporated into the H-reflex arc. The ischemic occlusion of the lowest-threshold afferents (more sensitive to ischemia) is offset for some time by the strong stimulus, which activates the higher-threshold (thinner) H-reflex afferents.

The ischemic occlusion of the thickest (fast-conducting) afferents, accompanied by unimpaired conduction along the thinner (slower-conducting) afferents, also explains why there is an increase in afferent H-reflex time during the ischemic period. The records in Figs. 6B and C show that the latency of the H-response increases much more than that of the M-response during ischemia. Just prior to its disappearance, the latency of the H-response is 2 msec longer than at the start of ischemia, while the latency of the M-response during this time is only 0.25 msec longer (Figs. 6D,E). In this respect, our data agree with the results of the two experiments by Mayer and Mawdsley (1965), although these researchers discovered a 5- to 6-msec increase in the latency of the H-response prior to its disappearance and no change in the latency of the M-response. However, such a significant increase in the latency of the H-response may not be real, since the latency will appear to exhibit a strong increase in the records as the H-response diminishes if the records are made with insufficient amplification (cf. records in Figs. 6B and C).

3. MONOSYNAPTIC NATURE OF THE H-REFLEX

In identifying the H-reflex as monosynaptic, a study was made of its posttetanic potentiation (PTP), since it is known that the PTP of a monosynaptic reflex has characteristic features (Lloyd, 1949; Kostyuk, 1954; Eccles, J., 1964) that distinguish it from the PTP of a polysynaptic reflex (Wilson, V. J., 1955). The PTP of the H-reflex in man was first described by F. A. Hoffmann (1952), and was later studied by other researchers as well (Homma and Tateiwa, 1960; Hagbarth, 1962; Blom et al., 1964; Corrie and Hardin, 1964; Mayer and Mawdsley, 1965; Lance et al., 1966). It is extremely difficult to compare the results of these studies, because of differences in the strength, frequency, and duration of the conditioning (tetanizing) and test stimulations of the popliteal nerve. Many of these parameters were not even indicated in the published results. Corrie and Hardin (1964) and Lance et al. (1966) noted the extreme variability of the effect of PTP of the H-reflex and its dependence on the strength of the tetanizing and test stimuli. Also, judging by experimental data, the PTP of a reflex can depend on the frequency of evocation of the reflex following tetanization (Predtechenskaya, 1965).

To study the phenomenon of PTP of the H-reflex for the purpose of comparing it with the phenomenon of PTP of the monosynaptic reflex of Lloyd, we chose the following parameters for the tetanizing and test stimuli applied to the popliteal nerve. Tetanizing stimulus: frequency, 40 or 60/sec (low-frequency tetanization) and 400/sec (high-frequency tetanization); duration of tetanization, 20 sec; strength of tetanization, either subthreshold, which is near the threshold for evoking the H-response of the Sol during single pretetanic stimulation, or suprathreshold, which elicits a control H-response with an amplitude up to 30-50% of the maximal. Parameters of test stimulus: frequency, approximately 0.1 Hz; strength, sufficient to elicit the control H-response of the Sol with amplitude 30-60% of the maximum. The H-response was evoked at 10-sec intervals. In treating the results of the experiment, the average amplitude was calculated on the basis of 5-7 H-responses before tetanization and 3 H-responses at 30-sec intervals after tetanization.

In an overwhelming number of experiments, no changes are observed in the shape or amplitude of the M-response following low- or high-frequency subthreshold tetanization (Figs. 7A,B,D), but such changes are observed after low- or high-frequency suprathreshold

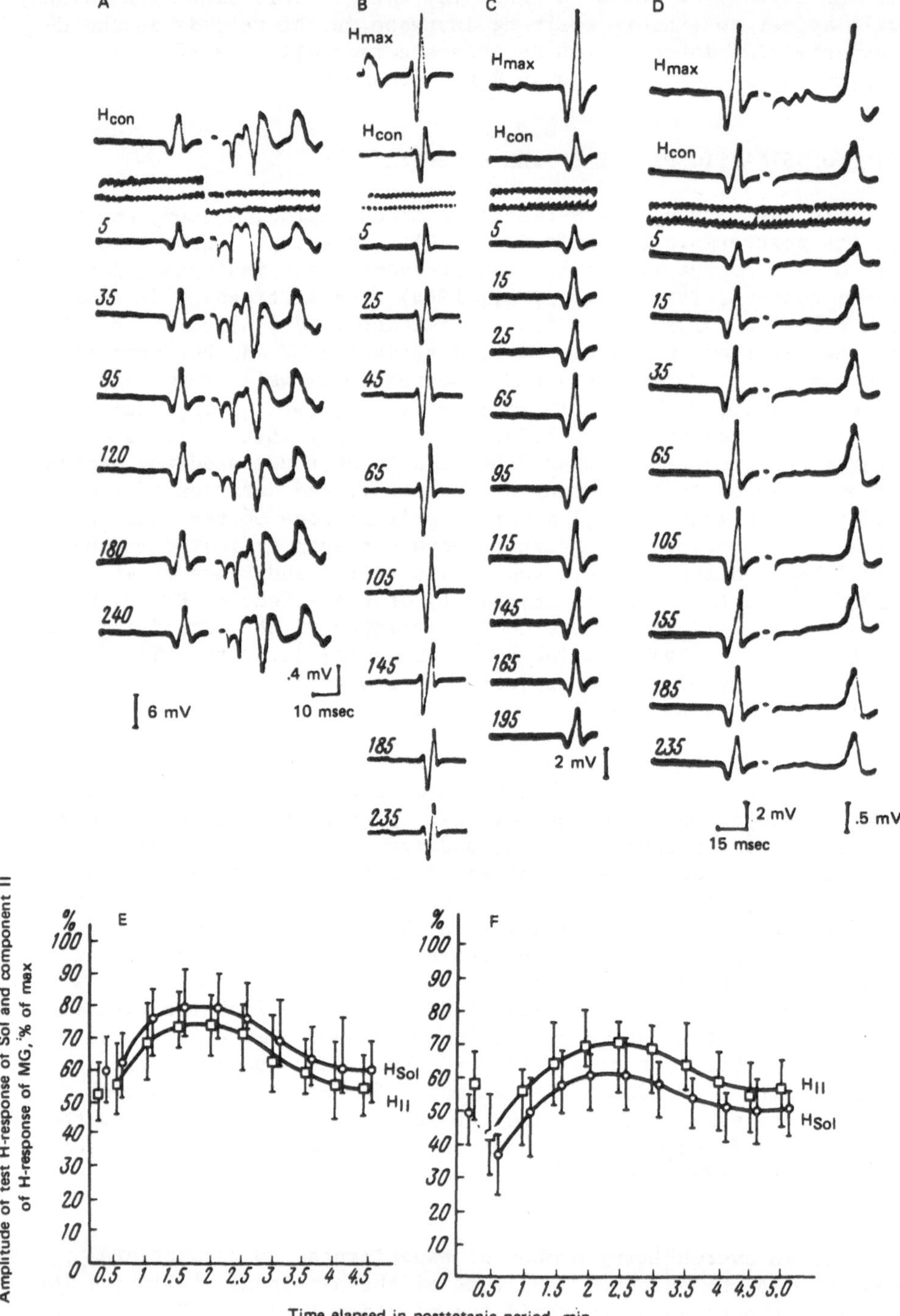
A
B
C
D
Hmax
Hcon
5
15
25
35
45
65
95
105
115
120
145
155
165
180
185
195
235
240
.4 mV
10 msec
6 mV
2 mV
15 msec
.5 mV
E
F
%
Amplitude of test H-response of Sol and component II of H-response of MG, % of max
HSol
HII
Time elapsed in posttetanic period, min

tetanization, which induces the contraction of the Sol. In some experiments, the M-response recovered its original amplitude and shape during the first 10-20 sec after tetanization, but in most cases, the changes in the M-response were irreversible. It is probable that no recoveries occurred when the stimulating and/or recording electrodes were dislodged by tetanic muscular contractions (Kots and Zaitsev, 1970). In analyzing the PTP of the H-reflex, the results of only those experiments in which the amplitude and shape of the M-responses did not change after tetanization were used.

After low-frequency, subthreshold tetanization, no changes are observed in the amplitude of the control H-response. Immediately after such tetanization, the amplitude of the H-response is either unchanged or (more frequently) below that of the pretetanic control response (see Fig. 7A). During the first 20-30 sec after tetanization, the H-response usually recovers its initial amplitude, which thereafter remains constant.

The basic changes in the H-reflex after high-frequency tetanization are shown in Figs. 7B-D, and E, F. During the first 10-20 sec following the cessation of high-frequency subthreshold tetanization, the amplitude of the control H-response is not noticeably changed, but it then increases over the next 1-1.5 min, rapidly at first, then more slowly (see Figs. 7B,E). Thereafter, there is a gradual decrease in the amplitude of the H-response to the original level. On the average (based on data pooled from 32 experiments), the maximum increase in the amplitude of the H-response of the Sol was 28.0±3.8%. The posttetanic effect usually persists for about 4-5 min (in some experiments, from 3 to 9 min).

After high-frequency suprathreshold tetanization, the amplitude of the H-response is, as a rule, less than the amplitude (pretetanic) for the first 10-20 sec (see Figs. 7C,D). The greater

---

Fig. 7. Posttetanic potentiation of the H-reflex. (A) H-response of Sol (left) and ATM (right) after low-frequency tetanization. (B) H-response of MG after subthreshold high-frequency tetanization. (C,D) H-response of Sol with amplitude ∼30% max. (C) and ∼50% (D, left) after suprathreshold high-frequency tetanization. (D, right) H-response of ATM. Black lines separate pretetanic (above) from posttetanic (below) records. Numbers beside records indicate time elapsed during posttetanic period in seconds. Upper records under B-D show maximal H-responses prior to tetanization. (E,F) Effects of high-frequency subthreshold (E) and suprathreshold (F) tetanization. Symbols at far left indicate control H-responses prior to tetanization.

the strength of the tetanizing stimulus, the greater the initial decrease in the amplitude of the control H-response (see Figs. 7C,D). After the depression period, the change in the H-response is similar to the changes following subthreshold high-frequency tetanization. In most cases, the maximum growth of the H-response is observed from the end of the 1st minute to the end of the 2nd minute of the posttetanic period. Based on data pooled from 27 experiments, the average maximum increase in the H-response of the Sol amounted to $36.2 \pm 4.2\%$ of the pretetanic control amplitude, i.e., somewhat greater than that observed after subthreshold high-frequency tetanization.

Thus, the PTP of the H-reflex in man exhibits features characteristic of the PTP of Lloyd's monosynaptic spinal reflex (Lloyd, 1949; Kostyuk, 1954; Wilson, V. J., 1955): (1) PTP of the H-reflex appears after high-frequency tetanization (400/sec), but not after relatively low-frequency tetanization (40 or 60/sec); (2) the maximum effect of PTP of the H-reflex does not appear immediately after tetanization, but several dozen seconds later; (3) the effect of PTP of the H-reflex generally lasts several minutes. Such a phenomenological similarity of the effect of PTP of the H-reflex to PTP of Lloyd's monosynaptic reflex tends to confirm the monosynaptic nature of the H-reflex in man.

That PTP of the H-reflex is observed at tetanizing stimulus strengths that are inadequate for the activation of motor fibers (i.e., no M-response) suggests the afferent "localization" of the phenomenon of PTP of the H-reflex (Eccles, J., 1964; Belekhova, 1968). This is in agreement with data obtained by Ström (1951) in acute experiments that revealed that tetanization of a muscle by stimulation of the distal portion of a sectioned ventral root does not provoke PTP of a monosynaptic test reflex evoked by stimulation of the gastrocnemius nerve. In addition, the appearance of PTP of the H-reflex after subthreshold tetanization suggests that those low-threshold afferents the stimulation of which during rest is inadequate to elicit an H-response are involved in the H-reflex.

The transient depression of the H-reflex immediately following suprathreshold tetanization cannot be accounted for by participation of the mechanism of the antidromic frequency activation of motoneurons, since it is known that such activation does not change a ventral root monosynaptic response elicited by stimulation of the dorsal root (Kostyuk, 1954). This phenomenon more likely stems from after-processes in motor nuclei activated reflexly during tetanization.

The data obtained by Magladery (1955) and our results provide additional evidence for the monosynaptic nature of the H-reflex in man. Moreover, experiments with tourniquet ischemia have supplied much new information on the afferent limb of the H-reflex arc. Judging by these data (see the preceding section), the H-reflex afferent limb is comprised of heterogenous afferent fibers. In addition to low-threshold, relatively thick (less resistant to ischemia) afferent fibers, the activation of which evokes suprathreshold H-responses from the Sol, the H-reflex afferent limb also appears to include higher-threshold, relatively thin (more resistant to ischemia) afferent fibers, the activation of which is required to elicit submaximal, maximal, and supramaximal H-responses from the Sol. This brings us to the question: to what group do these higher-threshold afferents belong, and what is the character of their central connections?

According to the scheme now accepted, only low-threshold I-a muscle afferents form direct, monosynaptic excitatory connections with the spinal motoneurons of their muscle. Also according to this scheme, I-b afferents from the Golgi tendon organs have a disynaptic inhibitory effect on the motoneurons of antagonist muscles (Eccles, J., 1957). Previously, it was believed that the relatively high-threshold Group II muscle afferents have a polysynaptic facilitatory effect on flexor motoneurons, largely independent of the muscle group (flexors or extensors) with which these afferents are connected (Eccles, R. M., and Lundberg, 1959b), and an inhibitory effect on extensor motoneurons (Holmquist and Lundberg, 1961). It was later shown, however, that Group II afferent fibers can have an excitatory effect on extensor motoneurons as well (Wilson, V. J., and Kato, 1965; Sverdlov, Yu. S., 1967).

Unfortunately, there are no experimental data available on the homonymous effects of II afferents on motoneurons in the absence of accompanying I-a afferent effects. An attempt was made to obtain such data in a study by Laporte and Bessou (1959). However, judging by the recordings presented in this work, the repeated high-frequency nerve stimulation used to block the thick fibers led to the development of a deep conduction blockade in Group II afferents as well as in Group I. According to the findings of Nakayama and Hori (1967), the relatively high-threshold afferents in the cat can apparently have a monosynaptic facilitatory effect on the spinal motoneurons of the GM.

The possibility of differences in the nature of suprathreshold H-responses elicited by the activation of relatively low-threshold afferents and that of near-maximal H-responses elicited by additional activation of relatively high-threshold afferents should be considered when choosing the magnitude of the test H-response.

One final comment should be made in connection with the use of the H-reflex method as a means of studying the state of spinal motoneurons in man. Following publication of the studies by Hoffmann and Lloyd, it was widely believed that testing over the elementary monosynaptic arc provided a direct measure of the excitability of spinal motoneurons. However, the discovery of presynaptic inhibition showed that the magnitude of the monosynaptic reflex depends not only on the level of motoneuronal excitability, but also on the state of presynaptic (afferent) pathways in the monosynaptic reflex arc. It is therefore more correct to say that the H-reflex method makes it possible to test the reflex excitability of spinal motoneurons, or the excitability of spinal motoneurons, through the monosynaptic reflex arc.

4. H-REFLEX OF THE SOLEUS AND MEDIAL GASTROCNEMIUS MUSCLES (DIFFERENTIATION OF "SLOW" AND "FAST" MOTONEURONS)

Studies in recent years have shown that the slow Sol and fast MG muscles in the cat are innervated by at least two functionally different alpha-motoneuron groups (Granit et al., 1956, 1957a,b; Eccles, J. C., et al., 1957, 1958, 1961; Kuno, 1959; Henneman et al., 1965a,b; Burke, 1968; Close, 1972; Granit, 1970). Granit and co-workers termed these two motoneuron groups "tonic" and "phasic"; Eccles called them the "slow" and "fast" groups, respectively; and Henneman and his colleagues designated them as "low-threshold" (small) and "high-threshold" (large). For Granit and co-workers, the basic criterion for such a classification was the character of the response of the motoneurons (duration of volley) to their reflex activation. It was shown in studies by Henatsch and Schulte (1958) and Kernell (1965b), however, that such a criterion is not very reliable: in different experimental situations, the response of the same motoneuron may vary from tonic to phasic, and vice versa. The classification suggested by Henneman and co-workers--low-threshold and high-threshold motoneurons--does not appear to be absolute, since the same population of motoneurons can apparently be low-threshold for activation through a different afferent input (see below). It is more accurate to speak of "slow" and "fast" motoneurons, and these are the terms that we prefer.

As shown in experiments performed by Granit, Eccles, and Henneman, the Sol of the cat is innervated chiefly (if not exclusively) by slow motoneurons, while the MG is innervated by both slow and fast motoneurons, the latter type making up most of the motoneurons of the MG.

The slow and fast motoneurons differ in the dimensions of their cell bodies. It was shown in a study by Burke (1968) that the Sol of the cat is innervated by small motoneurons (those with relatively high input resistance), whereas the MG is innervated by both small and large (with relatively high input resistance) motoneurons. The small motoneurons (with relatively high input resistance) have a relatively low axon conduction velocity, i.e., a thin axon, and the large motoneurons have a relatively high conduction velocity, i.e., a thicker axon.

The slow motoneurons of the Sol are characterized by a lengthier after-hyperpolarization than the fast motoneurons (Eccles, J. C., et al., 1958; Kuno, 1959), which is due largely to the fact that the frequency range in which the slow motoneurons can have stable function is lower than that of the fast motoneurons (Granit et al., 1956; Eccles, J. C., et al., 1958; Kernell, 1965a,b).

Judging by experimental data, some of the clearest criteria for a functional classification of Sol and MG motoneurons are the thresholds and levels of their reflex excitability on activation (electrical or muscle stretch) of the low-threshold Group I afferents: the slow Sol motoneurons have more effective monosynaptic connections with these afferents than the fast motoneurons of the MG (Eccles, J. C., et al., 1957; Kuno, 1959; Henneman et al., 1965a,b; Burke, 1968). Accordingly, the fraction of slow motoneurons in the Sol that are reflexly excitable by stimulation of the Group I afferents is greater than the fraction of MG motoneurons participating in the monosynaptic reflex (Denny-Brown, 1929; Kuno, 1959).

To compare the monosynaptic reflex excitability of Sol and MG motoneurons in man, we recorded the H-responses of the Sol and MG separately and determined the ratio of the amplitude of the maximal H-response to the amplitude of the maximal M-response (H_{max}/M_{max}) for each of these muscles (Kots and Krinskii, 1967). As noted earlier, the H-response of the Sol characteristically exhibits a triphasic potential, all phases of which change more or less uniformly (see Figs. 1A-D). The H-response of the MG (Fig. 8A) generally exhibits a more complex polyphasic potential, in which one can often discern two components--early (I) and late (II)--the amplitudes of which can undergo nonuniform changes under different conditions.* The data presented below tend to support the hypothesis that these two components of the MG H-response stem from the reflex activation of two types of MG motor units--the relatively high-threshold "fast" and low-threshold "slow" motor units.

Figure 8A shows the records of the H-responses of the MG, and the changes in the amplitudes of the H- and M- responses of the Sol and MG observed in one experiment are plotted in Figs. 8C and D.

---

*Two components cannot always be clearly discerned in the H-response. The reason for this is still unclear. It may be due to individual differences in the ratios of functionally diverse motor units (MU) in this muscle, their cross-sectional conformation in the muscle, the position of the branches of the medial gastrocnemius nerve, etc.

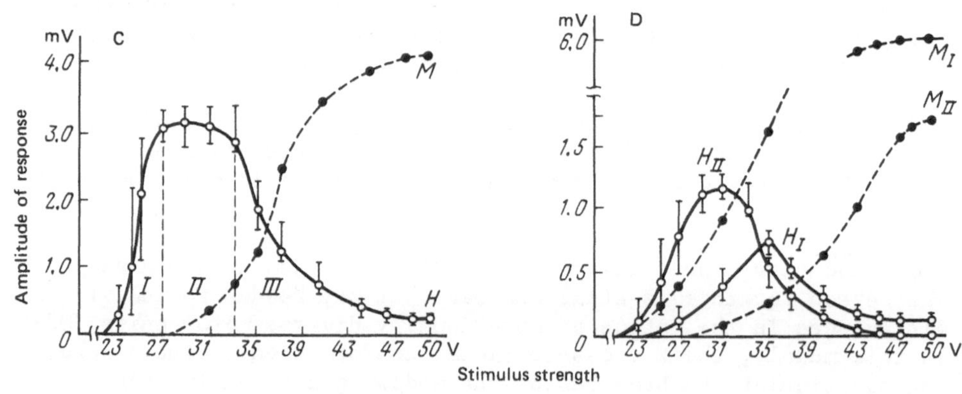
C
mV
4.0
3.0
2.0
1.0
0
Amplitude of response
M
H
I
II
III
23 27 31 35 39 43 47 50 V
D
mV
6.0
1.5
1.0
0.5
0
M_I
M_{II}
H_{II}
H_I
23 27 31 35 39 43 47 50 V
Stimulus strength

The Sol and MG generate threshold H-responses at equal or nearly equal stimulus strengths. During suprathreshold stimulation, the amplitude of component II of the MG H-response is several times greater than that of component I of the H-response of this muscle. As stimulus strength increases, the amplitudes of the Sol H-response and component II of the MG H-response also increase, though the amplitude of the H-response of the Sol increases somewhat more sharply. A somewhat greater stimulus strength is required to evoke the maximal amplitude of component II of the MG H-response than is needed to evoke the maximal H-response of the Sol (29 and 27 V, respectively, in our example). The amplitude of component I of the MG H-response rises most slowly as stimulus strength is increased, achieving a maximum at stimulus intensities at which the amplitudes of the Sol H-response and component II of the MG H-response begin to decline (35 V in our example).

The maximal amplitude of component I of the MG H-response is smaller (usually by a factor of 2-4) than the maximal amplitude of component II of the MG H-response. The maximal amplitude of component II of the MG H-response always has a lower voltage than the maximal Sol H-response. In this respect, our data are in agreement with data obtained by Levy (1963) and Mayer and Mawdsley (1965), who measured the amplitude of the MG H-response from peak to peak, i.e., by the greatest component II of the MG H-response. The ratio of the maximal Sol H-response to the maximal Sol M-response (H_{max}/M_{max}) fluctuated in our experiments from 50 to 90%, amounting to about 70% on the average. According to a communication by Táběikova (1966), this ratio may be as high as 100%.

At higher strengths of nerve stimulation (supramaximal), the Sol H-response and both components of the MG H-response decrease with increasing stimulus strength, and the H-response of the Sol and component II of the MG begin to diminish at a lower stimulus strength than component I of the MG H-response. As a rule, the H-responses are not completely abolished even at stimulus strengths that elicit the maximal M-responses of both muscles. At such strengths, component I of the MG H-response may have an amplitude up to 30% of the maximal, while the amplitude of the Sol H-

Fig. 8. H- and M-responses of the MG. (A) Responses are read from top to bottom of each column as the strength of nerve stimulation increases. The degree of amplification varies from one column to the next (see calibrations). The numbers beside the records indicate the strength of the stimulus in volts. (B) Records of H-responses of MG for interelectrode distances of 70 mm (left) and 5 mm (right). (C,D) Amplitudes of H- and M-responses of Sol (C) and two components of the H- and M-responses of the MG (D) as a function of the strength of nerve stimulation.

response and component II of the MG is very small. Component II is particularly depressed, and is almost absent in the presence of maximal M-response.

In most cases, the threshold of the MG M-response coincides with that of the H-response, or is even slightly lower. Usually, the suprathreshold M-responses of the MG exhibit a complicated pattern that cannot always be divided into two distinct components. The recordings in Fig. 8A made at different amplifications and the graph in Fig. 8D make it possible to trace the dependence of the amplitudes of the two components of the MG M-response on the strength of nerve stimulation. It can be seen that component I has the lower threshold of the two as it increases with a progressive increase in stimulus strength. The steepest portion of the component I growth curve corresponds to the sharpest initial decrease in the amplitudes of the Sol H-response and component II of the MG H-response. In this interval of stimulus-strength increase, the amplitude of component II of the MG M-response is comparatively small and rises relatively slowly. The maximum amplitude of component I of the MG M-response (for a 50-V stimulus) is approximately 4 times greater than the maximum amplitude of component II of the MG M-response. When measured from peak to peak--as was apparently done by Levy (1963)--the voltage of the maximum amplitude of the MG M-response is actually higher than that of the maximum amplitude of the Sol M-response.

How do we account for the time distribution of evoked electrical H- and M-responses of the MG? In experiments in which potentials were taken from the spinal cord during evocation of the H-reflex in man, Magladery et al. (1951) recorded the efferent (A) discharge in the form of a synchronous volley of 1- to 1.5-msec duration. When potentials are taken from the MG muscle, the H-response consists of two components with an interpeak interval of 4-5 msec; this interval is 1-1.5 msec shorter in the M-response. Therefore, the greatest "dispersion" of a response evidently occurs at the distal portion of the efferent limb, where there is a distal decrease in conduction velocity along the motor fibers (for a survey, see Baikushev et al., 1974). The dispersion of a response may depend only slightly on the small differences in conduction velocity along the membrane of different muscle fibers (Buchthal et al., 1955). In fact, changing the distance between recording electrodes from 1 to 70 mm effects no noticeable change in the interval between peaks in the M- or H-responses of the MG (Fig. 8B).

Component I of the M- and H-responses of the MG appears to be associated with the excitation of muscle fibers innervated by relatively fast-conducting efferent axons ("fast" motoneurons), and component II with the excitation of muscle fibers innervated by relatively slow-conducting axons ("slow" motoneurons).

Differences in the diameter of efferent axons explain why component I of the MG M-response appears at a lower stimulus strength than component II. Since the maximum amplitude of component I of the MG M-response is significantly greater than that of component II, we have evidence that the MG of the human, as well as the cat, contains more "fast" MUs than "slow" ones. However, it should be noted that in many muscles, fast MUs are larger and more superficial than slow MUs.

According to Lloyd (1943b), even a maximal antidromic stimulus applied to the ventral root more than 0.75 msec before stimulation of the dorsal root cannot completely block the monosynaptic reflex motoneuronal response. It is clear that if the anti- and orthodromic impulses are to "diverge," it would be most probable in the case of the fastest-conducting (low-threshold) efferent fibers. It is clear, therefore, why the reflex discharge of "fast" motoneurons of the MG (component I of the MG H-response) is depressed to a lesser degree by antidromic blockade during elicitation of the maximal M-response than the reflex response of the "slow" MG motoneurons (component II of the MG H-response).

The high conduction velocity along the fastest axons of the MG motoneurons (and possibly the Sol motoneurons) apparently enables the antidromic volley to exert (via these axons) a central inhibitory influence on the motoneurons of these muscles and thereby depress the H-reflex (see Section 1). In the medial gastrocnemius nerve of the cat, the fastest-conducting efferent axons have the same (or even higher) conduction velocities as the fastest-conducting afferent axons, amounting on the average to 108.6 and 105.8 m/sec, respectively (McDonald, 1963). Nor have differences in the conduction velocities of the fastest H-reflex efferents and afferents been detected in man (Mayer and Mawdsley, 1965). Considering the distance of impulse propagation, we may say that in man, an increase in the conduction velocity of the fastest-conducting efferent fibers over the conduction velocity of the bulk of the I-a afferents must lead to a difference of several milliseconds in the time of entry of the earliest antidromic and primary orthodromic (afferent) volleys into the spinal cord. This may be enough time for the development of the central (intraspinal) inhibition of the monosynaptic H-reflex owing to the activity of the Renshaw cells antidromically excited over the fast-conducting efferent axons.

According to data provided by Granit et al. (1957), Kuno (1959), and J. C. Eccles et al. (1961), in the cat, the greatest antidromic inhibition of the tonic motoneurons of the Sol takes place through recurrent collaterals of the axons of phasic MG motoneurons, while the latter generally display no antidromic inhibition. This selective central (recurrent) inhibition explains why the sharpest decline in the Sol H-response and component II of the

MG H-response occurs when, as stimulus strength is increased, there is a sharp rise in the amplitude of component I of the MG M-response accompanied by an increase in the amplitude of component I of the MG H-response. Clearly, as the strength of nerve stimulation increases, antidromic blockade in the efferent limb of the H-reflex begins to play an increasingly important role.

In experiments with cats, Granit et al. (1956) showed that the phenomenon of PTP is manifested in the facilitation of the stretch reflex responses of tonic ("slow") motoneurons, but that this phenomenon is absent in the case of phasic ("fast") motoneurons. In the light of these findings, we performed experiments (Kots and Zaitsev, 1970) in which posttetanic changes in the H-response of the Sol and component II of the H-response of the MG (reflex responses of low-threshold "slow" motoneurons) were compared with posttetanic changes in component I of the MG H-response (reflex response of relatively high-threshold "fast" motoneurons). According to our data, after low- or high-frequency subthreshold tetanization, the amplitude of component I of the MG H-response does not increase (Figs. 9A,B). After high-frequency tetanization with suprathreshold stimuli (evoking in the control an MG H-response with component I 50% of maximal), there is PTP of component I of the MG H-response (Fig. 9C), but the increase is considerably less than that of component II of the H-response of this muscle or the soleus. Thus, in our experiments (the records of which are shown in Fig. 9C) the maximum posttetanic increase in the amplitude of component II of the MG H-response was about 60% of the control amplitude, while that of component I was only 17%. Therefore, in experiments similar to those performed on cats (Granit et al., 1956), it was discovered that in man, the PTP of the reflex response of the low-threshold "slow" motoneurons (H-response of Sol and component II of H-response of MG) is substantially higher than the PTP of the reflex response of the high-threshold "fast" motoneurons (component I of H-response of MG).

Since our data and data in the literature indicate that facilitatory connections of "slow" and "fast" motoneurons of the GM with the low-threshold H-reflex afferents differ in their effectiveness, one might expect that the ischemic blockade of these fibers (see Section 2) would have different effects on changes in the reflex response of the two types of motoneurons. It has been pointed out--notably in a study by Vučo and Todorovic (1964)--that during compression of the femoral artery in the cat, the tonic motoneurons of the Sol lose their ability to fire continuously in response to a sustained muscle stretch, while the response of the phasic motoneurons of the MG does not change. In fact, it was found that component I of the MG H-response is less sensitive to ischemic deafferentation of the low-threshold H-reflex afferents than the Sol H-response and component II of the MG H-response.

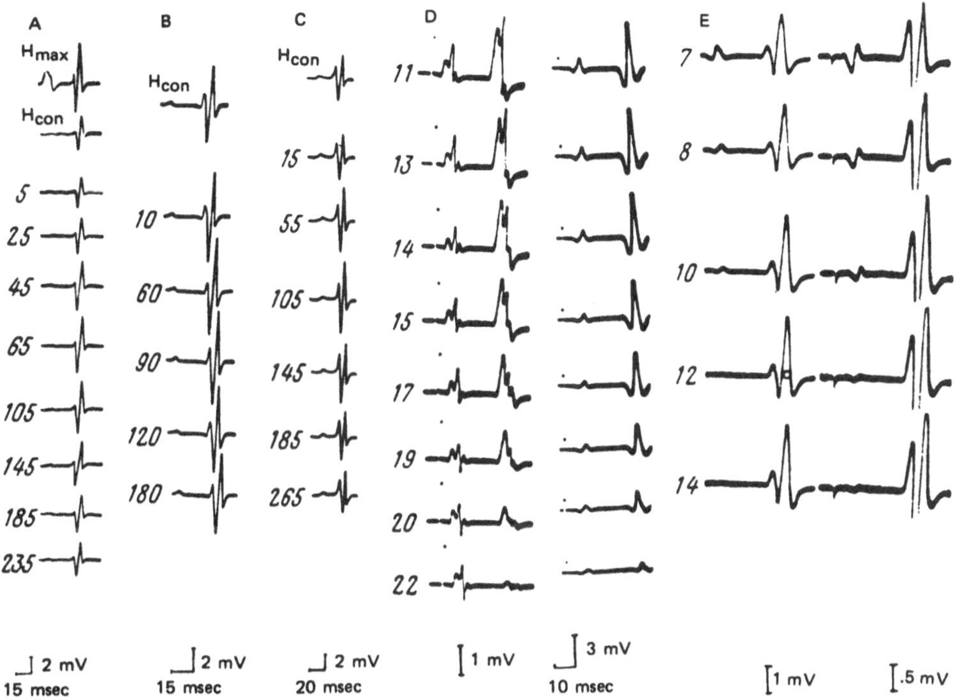

Fig. 9. Effect of high-frequency subthreshold (A,B) and suprathreshold (C) tetanization and of proximal (D) and distal (E) ischemia on the H-response of the MG muscle. (A) Control H-response of MG with amplitude of component I $\sim$15% and component II $\sim$50% of maximum. (B) Control H-response with amplitude of component I $\sim$45% and component II $\sim$70% of maximum. (C) Control H-response with amplitude of component I $\sim$40% of maximum. (D) H-responses of MG (left) and Sol (right) during proximal ischemia. (E) H-responses of MG (for different amplifications left and right) during distal ischemia. Numbers beside records indicate time elapsed after end of tetanization in seconds (A-C) and time elapsed after start of ischemia in minutes (D,E).

Figure 9D shows the records of the MG H-response with a submaximal amplitude of component I during proximal ischemia. For component II (and for the H-response of the Sol), such a stimulus strength is already supramaximal, so the amplitude of component II of the control H-response is depressed below its maximal amplitude. The records shown in the figure begin at the 11th minute of ischemia, when the amplitude of component II is just beginning to decay. Component I of the MG H-response declines later and more slowly. Thus, 14 min after the start of ischemia, the amplitude of component I is still unchanged, while that of component II is already reduced by 40%. At the 17th minute, the amplitude of component I is still about 90% of the control, whereas that of component II is already less than 30% of the control. Ischemic blockade of the low-threshold afferent fibers produced by a distal tourniquet leads to a considerably smaller increase in the reflex excitability of high-threshold "fast" motoneurons than of low-threshold "slow" motoneurons (Fig. 9E).

Thus, experiments with proximal and distal ischemia have shown that ischemic "deafferentation" has a greater effect on the reflex response of the low-threshold "slow" motoneurons of the Sol and MG than on the high-threshold "fast" motoneurons of the MG. We believe that the results obtained are best explained by the assumption that the reflex H-response of the relatively high-threshold "fast" motoneurons is largely dependent on the activation of afferent fibers that have a higher threshold of electrical stimulation and thus are thinner and more resistant to ischemia (Gasser and Erlanger, 1937; Fox and Kenmore, 1967) than the low-threshold afferent fibers the activation of which elicits an H-response from the "slow" motoneurons.

The presence of relatively high-threshold (i.e., thinner) afferents in the H-reflex arc of the "fast" motoneurons would account for the high resistance of the H-response of the "fast" motoneurons (component I of the MG H-response) to ischemia compared with the resistance of the H-response of the "slow" motoneurons (H-response of Sol and component II of MG H-response). If the H-reflex arc of the "fast" motoneurons does include relatively high-threshold afferents, we begin to understand why PTP of the "fast" motoneuronal reflex response (component I of MG H-response) is usually not evoked at all by subthreshold tetanization, and only to an insignificant degree by suprathreshold tetanization. It is possible that the tetanizing stimuli used in our experiments on PTP were inadequate to activate these afferents. Moreover, the phenomenon of PTP of spinal reflexes is known to be associated with changes in the tetanized presynaptic pathways (Lloyd, 1949; Kostyuk, 1954; Eccles, V. J., 1964).

There may be, however, another explanation for the slight influence of preliminary high-frequency tetanization on the reflex response of "fast" motoneurons. According to experimental data, high-frequency tetanization is followed by an increase of the monosynaptic spinal reflex, but not of the polysynaptic reflexes (Lloyd, 1949; Wilson, V. J., 1955). It remains unclear to which afferent group the relatively high-threshold H-reflex afferent fibers belong--I-a, I-b, or II--and what is the character of their intraspinal connections. If these high-threshold fibers have no monosynaptic connections with the "fast" motoneurons, it may account for the absence of high-frequency PTP of component I of the MG H-response.

In analyzing the nature of component I of the H-response of the MG muscle, we may draw on the experimental results of two studies. Kubota et al. (1965) recorded the neurogram of the tibial nerve (TN) and two of its peripheral branches--the MG nerve and the Sol nerve--in the cat in response to stimulation of the sciatic nerve. The reflex (H-) response of the TN is recorded as a two-peak volley with an interpeak interval of approximately 0.6 msec. Potentials recorded from the MG nerve display a single early peak, and from the Sol nerve, a single late peak. This led the authors to conclude that the reflex response of the MG motoneurons comprises the early component, and the reflex response of the Sol motoneurons the late component, of the two-peak response of the TN. It is significant that the interpeak interval in the TN response did not change as the stimulating electrodes were positioned nearer the spinal cord, the only result being a shortening of the latency of the reflex response. Thus, judging by the findings of this study, there appears to be no basis for assuming that the activation of the fast motoneurons of the MG requires the participation of higher-threshold (slower-conducting) afferents than the activation of the slow motoneurons of the Sol.

On the other hand, the findings of Nakayama and Hori (1967) are not consistent with such a conclusion. These authors also noted the complex shape of the H-response of the MG in man and performed special experiments with cats in order to analyze it. During stimulation of the ventral root in the muscle nerve of the MG and EMG activity of this muscle in the cat, a response with two peaks is again recorded: a lower-threshold, short-latency peak and a higher-threshold, long-latency peak. On stimulation of the central ending of the transected dorsal root, however, the long-latency component displayed a substantially lower threshold; a stimulus several times more intense than was adequate to evoke the long-latency component was required to evoke the short-latency component of the MG reflex response. According to these findings, therefore, the reflex activation of the fast motoneurons of the MG requires the participation of the relatively high-threshold dorsal root afferents. In addition, a comparison of the latencies of the short-latency components of the MG response during dorsal- and ventral-root stimulation indicates that these high-threshold afferents apparently possess monosynaptic excitatory connections with the fast motoneurons of the MG. Nakayama and Hori also found that after high-frequency tetanization (500/sec within 10-20 msec) in the cat, there is an increase only in the long-latency component of the H-response of the MG. On the basis of these findings and our own results, we are inclined to believe that the H-reflex arc of the "fast" motoneurons of the MG includes relatively high-threshold afferent fibers that apparently possess monosynaptic connections with these motoneurons.

On the basis of the results of studies of the Sol and MG H-reflex, the following conclusions may be drawn: As in other mammals, the motoneuronal pool of the Sol in man is comprised essentially of motoneurons that are relatively homogeneous in function and analogous to the "slow" motoneurons of the homonymous muscle in the cat. The motoneuronal pool of the MG includes at least two types of motoneurons, analogous to the "fast" and "slow" motoneurons of the homonymous muscle in the cat. The afferent limb of the H-reflex of the "fast" MG motoneurons includes higher-threshold (thinner) afferent fibers than the bulk of afferent fibers in the H-reflex arc of the "slow" Sol and MG motoneurons. The character of the central (intraspinal) connections between these high-threshold afferents and the motoneurons remains obscure.

CHAPTER 3

THE STATE OF SPINAL MOTONEURON AND INTERNEURON POOLS OF THE AGONIST DURING THE ORGANIZATION OF VOLUNTARY MOVEMENT

Morphological and electrophysiological studies have shown that the descending motor systems are connected with the spinal motoneurons either directly or through various interneuronal systems of the spinal cord. The direct connections between the descending systems and the motoneurons are generally assumed to consist only of monosynaptic contacts. In terms of function, however, we feel it is justified to include in this category connections through the specialized "interpolated" segmental interneuronal apparatus, which is selectively (or nearly selectively) activated by the given descending motor system and which relays its action directly to the motoneurons (Kostyuk, 1973).

Direct connections between descending systems and motoneurons have been investigated in detail in both the lower mammals (primarily cats) and the primates (Section 2), while the segmental interneuronal mechanisms of supraspinal influences have so far been studied only in cats (Section 4). The only exceptions are studies of the supraspinal effects on gamma-motoneurons (Section 3). In addition, histological and electrophysiological studies have shown that in both primates and cats, the terminals of the descending motor systems in the spinal cord are localized primarily in the interneuronal apparatus of the spinal cord. Regardless of which cerebral motor structures are involved in the organization of voluntary movement, and through which descending systems and by what supra- and intraspinal mechanisms their command actions are transmitted to the spinal motor centers, the end result of a voluntary motor command is a change in the state of the spinal motoneurons. Section 2 of this chapter describes the results of a study on the dynamics of the reflex excitability of the spinal motoneuronal pools of agonist muscles during the organization and at the onset of a phasic voluntary movement.

To analyze the dynamics of the state of spinal motoneurons during the organization of voluntary movement, it was important to ascertain (1) whether a supraspinal voluntary command produces changes in the state of the segmental interneuronal apparatus and, if so, what changes; (2) which interneuronal systems are subjected to supraspinal regulatory actions during the organization of voluntary movement; and (3) the time course of functional reorganizations in the interneuronal segmental apparatus during this period. In Sections 3 and 4, the results of a study on the dynamic state of the interneuronal apparatus of the agonist during the organization of voluntary movement are discussed. One portion of this study was aimed at identifying the role of the supraspinal activation of fusimotor (gamma) motoneurons in the initiation of voluntary movement (Section 3). In terms of function, the gamma-motoneurons may be regarded as interneurons with a facilitatory effect (via the gamma loop) on the homonymous alpha-motoneurons. The supraspinal activation of gamma-motoneurons may be used as one mechanism for the facilitation of the motoneuronal pool of the agonist of a voluntary movement. Another portion of the study dealt with the dynamic state of the spinal inhibitory interneuronal apparatus associated with this pool (Section 4).

1. EFFECT OF THE H-REFLEX ON THE LATENT PERIOD OF A VOLUNTARY MOVEMENT

Since the H-reflex (or Achilles reflex) was used in our studies to test the state of the spinal motoneuronal pool of a future agonist, we must ask whether elicitation of the test reflex influences (and if so, how) the onset of voluntary impulse activity of the agonist motoneurons. It is known that the synchronized motoneuronal discharge accompanying elicitation of the H-reflex is followed (as shown in experiments with paired stimuli) by a transient period of facilitation lasting no longer than a few milliseconds (Magladery et al., 1951b), and then by a longer period (up to several hundred milliseconds) of depressed motoneuronal excitability (see Section 4). It can therefore be assumed that the test H-reflex will affect the initiation of voluntary impulse activity of the agonist motoneuronal pool, accelerating or delaying the appearance of voluntary activity, depending on the moment at which the test H-reflex is evoked.

The solution to this problem depends primarily on the method of measuring the interval between the moment at which the H-reflex is evoked and the onset of voluntary movement. In addition to methodological aspects, the solution to this problem is also important in identifying the characteristics of the interaction of supraspinal (voluntary) and evoked segmental reflex effects on the spinal motoneuronal pool of the future agonist in question.

To determine the effect of the test reflex on the latency of voluntary movement, we performed a series of special experiments. In every case, the voluntary movement consisted of a rapid extension of the foot (lifting heel from floor) in response to a movement signal. Movements performed in response only to the signal were alternated with movements in which the H-reflex was evoked either simultaneously with or slightly after presentation of the signal, but before the onset of the voluntary movement. Then the latent periods obtained from these two variations were compared. The voluntary EMGs of all three heads of triceps surae--the medial gastrocnemius, lateral gastrocnemius, and soleus muscles-- began simultaneously during the chosen phasic movement. The onset of voluntary movement was usually read from the EMG of the MG, which was recorded at high amplication, while the H-response of the Sol was used to measure the amplitude of the H-response, since its simple shape permitted a more precise measurement.

In the first series of experiments (V. Ivlev), a maximal test stimulus was used, and this stimulus evoked a maximal H-response during rest. This stimulus was applied concomitantly with the signal (light flash) and 40, 80, and 110 msec after presentation of the signal. In each variation, the movement was repeated 5 times, the variations being randomly alternated throughout the experiment.

As our data indicate (Fig. 10A), when the stimulus evoking the maximal H-response is applied concomitantly with the signal, the latent period of the voluntary movement is significantly shorter than that of the other cases. If the stimulus is applied after the signal, motor reaction time remains constant or becomes somewhat longer than reaction time in response to a signal only (without test stimulus). The latent period tends to increase in proportion to the length of the delay from signal presentation to evocation of the H-reflex. For example, for a 40-msec delay, average motor reaction time is 4.5 msec longer than reaction time in response to a signal only; delays of 80 and 110 msec increase average reaction time by 10.9 and 15.3 msec, respectively. The difference does not appear to be statistically significant in any of these cases, however. As a comparison shows, increasing the delay by 40 msec--from 40 to 80 msec--lengthens reaction time by an average of only 6.4 msec, and an additional increase of 30 msec--from 80 to 110 msec--by only 5.4 msec. Thus, a total increase of the delay by 70 msec--from 40 to 110 msec--lengthens reaction time by only 11.8 msec. We may therefore conclude that if the reflex discharge of a significant number of the motoneurons of the future voluntary agonist does have an effect on the onset of voluntary impulse activity by the motoneuronal pool of the muscle in question, this effect is not significant.

The slight lengthening of reaction time observed may be explained as follows. As the delay increases, the H-response appears more and more often in intervals immediately preceding the onset of voluntary impulse activity by the motoneurons, or coincides with

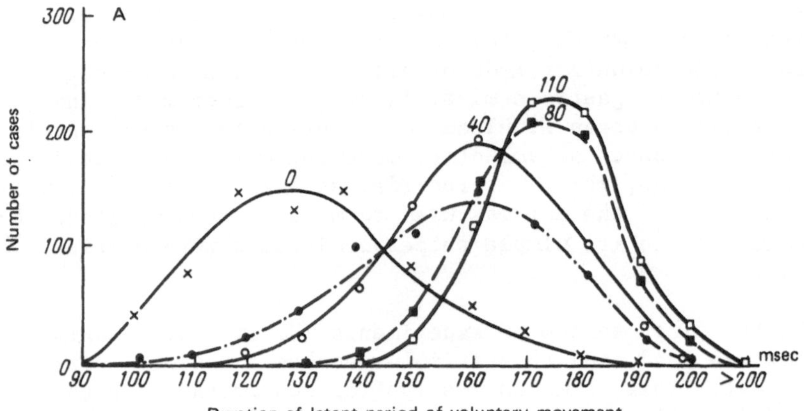

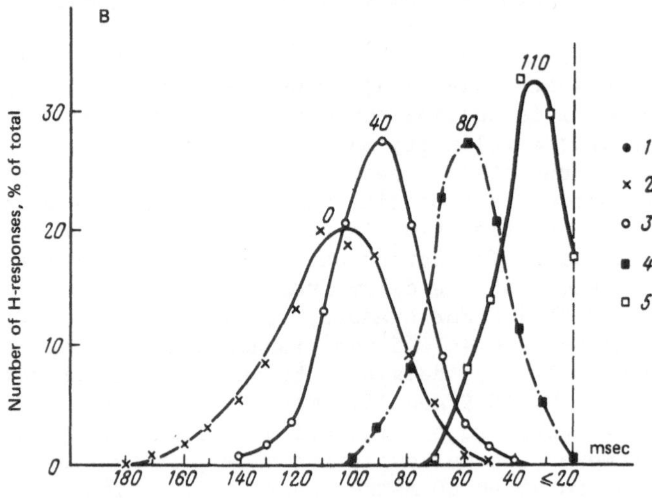

Fig. 10. Distribution of latent periods of voluntary movement for different delays between visual signal presentation and nerve stimulation (A) and distribution of number of H-responses in various time intervals prior to movement (B). (1) Without testing (2-5) during testing with different delays: (2) zero, (3) 40 msec, (4) 80 msec, (5) 110 msec.

the onset of this activity (Fig. 10B). This suggests the following possibilities, all of which may affect the measurement of average motor reaction time. First, if the EMG volley of voluntary agonist activity immediately follows the H-response, it is impossible to determine whether the voluntary EMG activity began immediately after the end of the H-response, or during the H-response, or accompanied the onset of the H-response. Considering that the duration of the H-response of the GM is approximately 15 msec, it can be seen that in the cases described above, the true onset of voluntary activity cannot be determined with an accuracy better than ± 15 msec. Second, the reflex firing of motoneurons on evocation of the H-reflex can lead to a brief period of motoneuronal inactivity (refractoriness), which will, of course, delay the onset of voluntary EMG activity. That this delay is very small (several milliseconds), even after evocation of the maximal H-reflex, permits the following assumptions to be made: (1) during the period preceding the onset of voluntary movement, the refractoriness of the discharge motoneurons is quite short, permitting their repetitive discharge on voluntary activation after a very short interval of time; (2) the initial portion of the voluntary volley comprises the impulses of those motoneurons not involved in the reflex discharge. The first of these assumptions seems quite probable, since even the H-response equal to M_{max}, i.e., that which recruits practically all the motoneurons into the reflex discharge, can be followed by voluntary EMG activity of the agonist (see the next section).

The absence of a significant effect of the agonist H-reflex on the onset of agonist voluntary activity is also indicated by data on the frequency distribution of H-responses over different intervals prior to voluntary movement. These curves have a normal, symmetrical shape for all delays between signal presentation and the evocation of the maximal H-reflex (0, 40, 80, and 110 msec). Increasing the delay simply displaces the curve toward the onset of voluntary EMG activity with no change in the shape of the curve. No characteristic intervals between the H-response and the onset of voluntary activity have been observed, nor has the absence of an interval, as would be expected if the H-reflex had a significant influence on the onset of voluntary motoneuronal impulse activity.

However, when the maximal H-reflex is evoked by nerve stimulation that accompanies signal presentation, there is a considerable reduction in the latency of voluntary movement. It is likely that in this case the electrical nerve stimulation is transformed into a movement signal, since it is known that reaction time is shorter for tactile and proprioceptive signals than for visual stimuli (Boiko, 1964). When the stimulus that evokes the H-reflex is applied after the visual signal is presented, the subject apparently reacts to the earlier signal--the light flash. This explains why the average reaction time to a visual signal is statistically equivalent for different delays in nerve stimulation after the signal is presented. It is also possible that the shortened reaction time accompanying simultaneous signal presentation and nerve stimulation is due in part to an overall amplification of the movement signal, which can also lead to a reduction in reaction time (Boiko, 1964; Avakyan et al., 1966).

In the second and third series of experiments (Mart'yanov, 1968), the subjects performed a movement in response to a signal, followed by nerve stimulation that elicited a resting, suprathreshold H-response with an amplitude 30-50% of the maximal. The delays between signal presentation and the test reflex were selected such that the H-response usually appeared in the interval approximately 10-90 msec prior to the voluntary EMG volley. The choice of such delays was determined, on the one hand, by the consideration that if the evoked synchronized response of the motoneurons affects the onset of their voluntary impulse activity, then apparently the effect is stronger the nearer the appearance of the response is to the onset of the voluntary movement. This appears likely, since, as will be shown below, the number of motoneurons that respond to equal stimulus strengths is greater, the nearer the H-reflex is evoked to the onset of movement. On the other hand, when the testing stimulus is applied to the nerve simultaneously with or slightly later than the presentation of the movement signal, the possibility that this stimulus is transformed into a movement signal cannot be discounted.

In all 13 subjects participating in the second series of experiments, no differences were found between average motor reaction time in response to a visual stimulus obtained without evocation of the H-reflex and with evocation of this reflex during the last 100 msec of the latent period. In these experiments, the average intervals from the test H-response to the appearance of the voluntary EMG volley of the agonist were distributed as follows:

Cases, %	Intervals, msec
18.6	0-29
24.9	30-49
27.9	50-69
21.7	70-89
6.9	>90

Analogous results were obtained during the third series of experiments, in which electrodermal stimulation of the earlobe served as the movement signal. In these experiments, the average interval from the test H-response to the onset of voluntary EMG activity was within the following limits:

Cases, %	Intervals, msec
22.5	30-49
27.0	50-69
22.5	70-89
22.5	90-100
5.5	>100

The average difference in the duration of the "control" and "tested" latent periods amounted to 0.08 msec.

Thus, during at least the last 100 msec of the latent period, evocation of the H-reflex has no effect on the onset of the voluntary EMG volley. The absence of this effect makes it possible to determine the moment of evocation of the test H-reflex by the time interval from the H-response to the onset of voluntary EMG activity by the agonist. The mechanism of the surprising tolerance of the initial voluntary impulse activity of the motoneuronal nucleus to the action of the evoked reflex will be discussed in Section 4.

2. THE MOTONEURON POOL OF THE AGONIST

In primates, the cortical motor areas have direct (monosynaptic) connections with spinal motoneurons (Hoff, E. C., and Hoff, 1934; Kuypers, 1960; Liu and Chambers, 1964). The relative number of these monosynaptic corticomotoneuronal connections increases from the lower monkeys up through the higher primates and man (Kuypers, 1960, 1964). According to histological data, the corticospinal (pyramidal) pathways form the densest monosynaptic contacts with motoneurons located in the dorsolateral area of the ventral horn (Kuypers, 1964; Liu and Chambers, 1964). The motoneurons of this area innervate primarily distal limb musculature (Sharrard, 1955; Rexed, 1964; Romanes, 1964). In monkeys, the fastest corticospinal fibers have direct excitatory connections with the spinal motoneurons of the flexors (Bernhard et al., 1953; Preston and Whitlock, 1960, 1961; Agnew et al., 1963; Laursen and Wiesendanger, 1966; Stewart and Preston, 1967). In the motoneurons of the upper limbs, these monosynaptic connections are more numerous than in the motoneurons of the lower limbs. The motoneurons of muscles of the distal members also have more numerous direct corticospinal connections than those of the muscles of more proximal members (Bernhard and Bohm, 1954). Since these muscles are "represented" primarily in the anterior wall of the central sulcus (Woolsey, 1958), and since the Betz cells are located chiefly in this area (Lassek, 1954), many authors have quite naturally assumed that the thick, fast-conducting axons of the Betz cells constitute most, if not all, of the axons that are monosynaptically connected with the motoneurons.

At the same time, it is known that thick, fast-conducting axons constitute only a very small part of the pyramidal fibers (Lassek, 1954). Thus, in determining the functional role of the monosynaptic corticomotoneuronal connections of the thick pyramidal fibers, one must consider data on their effectiveness in motoneuronal activation. According to data published by Phillips and Porter (1964), brief, high-frequency (200/sec) stimulation of pyramidal tract axons produces a progressive increase in the depolarization of motoneuronal membranes: the depolarization level increases by a factor of 10 over that induced by the first stimulus. Phillips and Porter believe that such a powerful iterative effect of corticospinal activity must ensure the rapid initiation of movements. During natural movements, however, the pyramidal tract neurons do not have such a high discharge frequency (see Chapter 1). When a low-frequency--about 25/sec--stimulus is applied to the motor cortex, several seconds are required for direct (monosynaptic) motoneuronal excitation (Bernhard et al., 1953). Thus, even for the synchronous stimulation of a large number of descending pyramidal fibers, indirect polysynaptic motoneuronal excitation appears to be more effective than the direct monosynaptic pathway (Kostyuk, 1973). Kostyuk believes that under conditions of natural activity, the segmental and suprasegmental afferent inputs set up a depolarization "background" in the motoneurons, which can then be fired by even very-low-amplitude pyramidal excitatory postsynaptic potentials (EPSPs), or effectively inhibited by inhibitory postsynaptic potentials (IPSPs).

Until quite recently, monosynaptic corticomotoneuronal connections in primates were thought to be a unique property of the pyramidal tract (Wiesendanger, 1969). These connections were believed to play a predominant role in the control of fine limb movements (Bernhard et al., 1953).

It has been determined in a number of recent studies, however, that besides the corticospinal (pyramidal) system, other (extrapyramidal) descending motor systems in monkeys and other mammals also form monosynaptic excitatory connections with spinal motoneurons (Shapovalov, 1968, 1970, 1972). Monosynaptic EPSPs were recorded only in single motoneurons during stimulation of the red nucleus (Shapovalov and Shapovalova, 1966; Shapovalov and Karamyan, 1968; Shapovalov, 1972; Hongo et al., 1969a), in a number of (for the most part extensoral) motoneurons during activation of the vestibulospinal pathways (Shapovalov et al., 1966; Lund and Pompeiano, 1968; Shapovalov and Saf'yants, 1968; Grillner and Lund, 1968; Shapovalov, 1972), and in motoneurons (primarily flexoral, but often extensoral) on activation of the reticulospinal pathways (Shapovalov et al., 1967; Grillner and Lund, 1968; Shapovalov, 1972).

Thus, monosynaptic excitatory connections with spinal motoneurons appear to be a property shared by various descending motor systems. Shapovalov (1970) points out some basic common features that characterize the monosynaptic connections of the pyramidal and extrapyramidal descending systems with the spinal motoneurons. First, it is characteristic of any descending motor system that its fastest-conducting (i.e., thickest) fibers are formed by monosynaptic contacts with the motoneurons. Such fibers constitute only a small fraction of the total fiber mass of descending systems. Second, the effectiveness of supraspinal monosynaptic actions on motoneurons is relatively low, such that in themselves these actions do not induce motoneuronal excitation. The effectiveness decreases caudally, such that monosynaptic facilitation of lumbar motoneurons via extrapyramidal pathways from the stem structures (Wilson, V. J., and Yoshida, 1969) or via the pyramidal pathway from the motor cortex is weaker than that of motoneurons of the cervical enlargement. These findings led Shapovalov to propose that the functional role of descending monosynaptic effects is largely determined by their interaction with other synaptic inputs to the spinal motoneurons. By tracing the phylo- and ontogenetic development of descending monosynaptic connections with the motoneurons, Shapovalov came to the conclusion that the evolutionary cephalization of the nervous system has been accompanied by an increase in the number of direct connections from the brain to the spinal motoneurons, and by a shift in the origin of these connections to the motor area of the cerebral cortex. Evolutionary developments are also reflected in an increasing differentiation of the monosynaptic connections of different supraspinal motor centers with the motoneurons of functionally diverse muscle groups.

While studying the distribution of focal potentials in the lumbar enlargement of the cat spinal cord, Vasilenko and Kostyuk (1965, 1966) found that pyramidal input leads to the excitation of relatively specialized interneuronal groups, quite distinct from interneurons of the reflex arc. One such group is located in the external basilar region, where the corticospinal fibers terminate. The neurons of this group are monosynaptically excited by the fast-conducting pyramidal fibers, and not by impulses from peripheral afferent fibers. Based on the effects of antidromic stimulation of the lateral cord, these neurons were identified as short propriospinal neurons. Their axons extend caudally for two or three segments and can apparently have a monosynaptic influence on motoneurons within the limits of these segments. The effect of the propriospinal neurons of the external basilar region coincides with that observed during stimulation of the pyramidal tract--the generation of EPSPs primarily in the flexoral and IPSPs in the extensoral motoneurons (Kostyuk et al., 1969). It is therefore quite likely, as Kostyuk suggests, that the external basilar neurons are elements of a fast-conducting, specialized system "interpolated" between the fast-conducting pyramidal fibers and the motoneurons. In the absence of monosynaptic corticomotoneuronal connections, pyramidal actions in the cat are transmitted to the motoneurons via these specialized propriospinal interneurons.

It is not yet known whether there is a similar specialized interneuronal "interpolated" system for the corticospinal pathways in primates, but the existence of such a system is entirely possible. For example, according to data obtained by Preston and his associates (Uemura and Preston, 1965; Stewart and Preston, 1967), the interneuronal pools that determine the late cortically evoked facilitation of spinal neurons require the summation of repetitive pyramidal discharges to achieve the maximum effect, similar to the interneurons of polysynaptic corticospinal systems in the cat (Lloyd, 1941b; Hern et al., 1962).

In addition to the corticospinal (pyramidal) systems, it is likely that there are other descending motor systems that can influence the spinal motoneurons via specialized "interpolated" interneuronal systems. According to electrophysiological data, the fibers of descending tracts of the lateral columns in the cat--the rubrospinal and parts of the reticulospinal--terminate in isolated groups of spinal interneurons (Sasaki and Tanaka, 1964; Hongo et al., 1965, 1969; Kostyuk, 1967, 1970; Kostyuk and Pilyavskii, 1969; Pilyavskii and Skibo, 1969). These interneurons are monosynaptically activated by impulses from only these descending pathways, and do not respond to peripheral nerve stimulation. The cell bodies ot the specialized rubrospinal interneurons are located in the lateral portion of lamina VII (according to Rexed), which is primarily the focal point of rubrospinal fiber endings. These interneurons probably belong to the group of short, propriospinal neurons that establish monosynaptic connections with spinal alpha-motoneurons (Kostyuk and Pilyavskii, 1969; Kostyuk, 1970, 1973). The influence of this interneuronal system on other spinal interneurons is insignificant.

In experiments performed by Erulkar et al. (1966) involving the stimulation of the vestibular nerve in the cat, the presence of segmental interneurons that are selectively activated by this stimulation and do not react to other descending (including corticospinal) and afferent influences was established. The localization of these neurons (lamina VIII and the medial portion of lamina XI) coincides with the region in which primarily vestibulospinal fibers terminate.

Thus, there are several descending motor systems that can excite spinal motoneurons through direct (monosynaptic) connections with the spinal motoneurons or through specialized (and other) interneuronal segmental systems. However, it is still not known which of these supraspinal descending and interneuronal segmental motor systems participate in the activation of spinal motoneuronal pools during the organization and regulation of voluntary (and other natural) movements, and what the mechanisms of this participation are.

In our studies on the dynamics of the state of agonist motoneurons during the organization of voluntary movement, we applied the monosynaptic testing method (H-reflex). The movement signals consisted of a faint click presented through headphones, a neon light flash, or electrical stimulation of the earlobe. It was determined in special experiments that if these stimuli are not recognized as movement signals, they do not produce significant changes in the H-response (see Chapter 5, Section 1). Like other authors (Boiko, 1964), we discovered that average reaction time is influenced by the modality of the stimulus. On the other hand, since the nature of changes in the H-response during the latent period of voluntary movement is not dependent on signal modality, the results of experiments that employed different movement signals will be discussed collectively. To reduce the duration and variability of the latent period, a "warning" signal was given prior to the movement signal (Chuprikova, 1967).

The subject was instructed to respond to the movement signal by lifting his heel from the floor as rapidly as possible (extension, or plantar flexion, of the foot). After various post- or presignal delays, a test H-reflex of the GM-extensors of the foot was evoked. Throughout the experiment, the signaled movement was repeated at least 100 times, and the delays in evoking the H-reflex were varied such that the H-response of the agonist would appear at different intervals prior to the onset of voluntary movement. We were thus able to trace changes in the test H-response over the entire latent period. The intervals between repetitive movements were varied during the experiment from 8 to 15 sec. Such intervals minimized the effect of the test stimulus on the amplitude of the subsequent H-response (see Section 4). Throughout the experiment, the amplitude of the test H-response was continuously monitored during the rest periods between movements, and the stimulating and recording conditions were kept constant by monitoring the shapes and amplitudes of the M-responses.

As has been shown in experiments designed to measure the reflex excitability of the agonist motoneuronal pool, the latent period of a voluntary movement can be divided into three phases, which we have called the "pretuning," "tuning," and "triggering" phases. The first phase, the "pretuning," begins prior to presentation of the movement signal; the second phase, the "tuning," appears 55-60 msec prior to voluntary movement; the third phase, the "triggering," begins 25-30 msec prior to movement (voluntary impulse activity of the agonist motoneuronal pool).

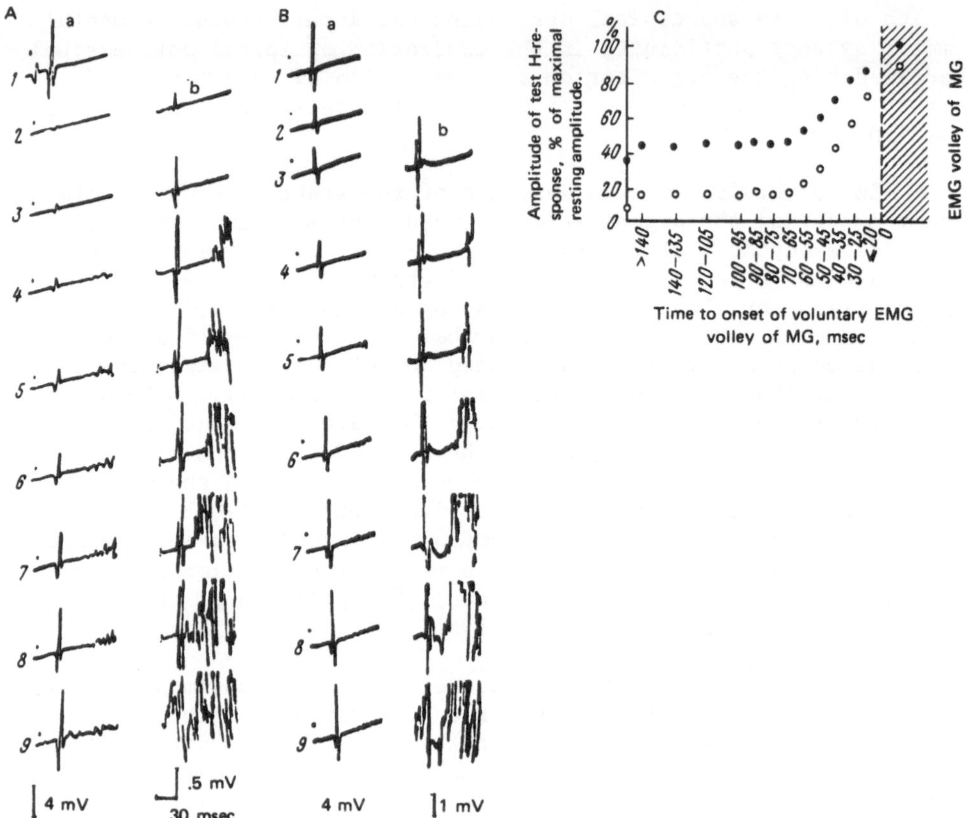

Fig. 11. H-response of Sol during latent period and at onset of voluntary foot extension. A,B: (a) EMG of Sol; (b) EMG of MG. (1) Maximal H-response; (2) control H-response, resting; (3-7) H-responses nearing onset of voluntary EMG volley of MG; (8,9) H-response at onset (8) and after onset (9) of EMG volley of MG. (A) Near-threshold; (B) suprathreshold test H-responses of Sol; (C) change in test H-response of Sol during latent period of voluntary movement. Open circles represent H-response at rest.

The Pretuning Phase

The amplitude of the test H-response of the agonist is appreciably greater at the onset of the latent period than during rest (Fig. 11). This early increase in the reflex excitability of the motoneuronal pool of the future agonist is not induced by the movement signal, since it appears prior to signal presentation. It remains constant throughout the early phase of the latent period, which precedes the onset of movement by more than 60 msec. We have termed this increase in reflex excitability the "pretuning" increase. The reasons for this designation will be discussed presently. It is noteworthy that the relative increase in the amplitude of the H-response during the first phase of the latent period is inversely proportional to its initial control amplitude (see solid and open circles in Fig. 11C).

An increase in the amplitude of the test H-response is recorded not only during the initial phase of the latent period of foot extension, in which the foot extensors (GM) serve as the primary agonists, but also during the latent period of quite different voluntary movements, in which these muscles either serve as direct antagonists (dorsal flexion of the foot), are sympathetically activated or inhibited (during voluntary extensions or flexions in the hip or knee joint of the test limb), or are not activated at all, e.g., during movements of the contralateral leg or of the ipsi- or contralateral hand. Throughout most or all of the latent period of these movements, the increased (over resting) amplitude of the test H-reflex exhibits no noticeable change. As in the case of extension of the foot, an increase in the test H-response of the GM can be observed both after the movement signal and during the period of signal expectancy.

Thus, the increase in the reflex excitability of the spinal motoneurons of a given muscle--an increase that appears prior to the movement signal and persists for most or all of the latent period--appears to occur before any movement, regardless of the function of this muscle in the pending movement. At one time, these facts prompted us to regard the increased (over resting) reflex excitability observed during the first phase of the latent period as the product of a general "nonspecific" attention reaction preceding the movement signal, of a general "readiness" to move, independent of the type of movement pending. In this sense, such increased motoneuronal excitability was identified as "nonspecific," and the first phase of the latent period of voluntary extension of the foot (characterized by a stable plateau value of heightened motoneuronal excitability in the future agonist) was termed the "nonspecific" phase (Gurfinkel' and Kots, 1966).

It was shown in further studies, however, that--like other obvious motor or electromyographic manifestations of the "ideomotor reaction" (Bassin and Serkova, 1956; Bassin and Sidorov, 1963; Vinogradov and Shvang, 1957; Kolodnaya, 1957)--early changes in the reflex excitability of the motoneuronal pool do not appear to be "nonspecific," but rather definitely related to the movement to be performed, i.e., to the function of the muscle innervated by the given motoneuronal pool in this movement. Such a conclusion is particularly supported by the results of experiments in which subjects performed various movements in response to a light stimulus: ventral flexion of the right foot of the tested leg (1),

Table 1. Average Amplitudes of H-Responses at Onset of Latent Period of Different Movements (in % of Control Amplitude at Rest)

	Movement					
Index	1	2	3	4	5	6
χ	123.2	112.5	112.0	112.7	113.5	113.4
σ	±14.3	±11.2	±18.6	±15.1	± 7.2	±10.8
Number of experiments	21	11	14	11	11	11

of the left (nontested) leg (2), dorsal flexion of the right (nontested) foot (3) and flexion in the elbow joint of the right (ipsilateral) hand (4). A special series of tests involved a choice, whereby the subject was to extend the right (tested) foot (5) in response to illumination of the right lamp and extend the left (nontested) foot (6) in response to illumination of the left lamp. In each of these cases, the H-response was elicited in the right leg simultaneously with the presentation of the movement signal. In each such experiment, the movement with the test H-reflex was repeated 30 times, and the control H-response was elicited 20 times (10 times at the start of the experiment and 10 times at the end).

The results of the experiments (Table 1) showed that even an early increase in the reflex excitability of motor nuclei is quite specific. Besides the motoneuronal pools of the future agonists of the voluntary movement, the state of signal expectancy, i.e., the readiness to perform a particular movement in response to this signal, is accompanied by a roughly identical increase in the reflex excitability of all remaining motoneuronal pools. This "background" ("diffuse," "nonspecific") increase in excitability is apparently one of the physiological correlates of the state of signal expectancy. According to the findings of Ovsyannikov and Khomyakova (1969), the more complex the motor problem, the greater the early increase in the H-response ("background," according to our terminology). Tzekov (1972) also found a correlation between reaction time and H-response amplitude during the readiness period.

Standing out in the midst of this "background" excitability increase is a more significant, "specific" increase in reflex excitability, which is limited to the motoneuronal pools of the agonists involved in the future voluntary movement. It is thought that the phenomenon of a selective increase in the excitability of the motoneuronal pools of the future agonists may be regarded as one of the physiological manifestations of adjustment (presetting), i.e., of a special psychophysiological readiness state for a particular activity (Uznadze, 1961).

According to a hypothesis formed by Bernstein (1961), the onset of movement is preceded by a preliminary "tuning" of the excitability of all participating sensory and motor elements, this tuning being performed by the central nervous system in accordance with the intended motor program. This preliminary tuning includes a selective increase in the reflex excitability of the motoneuronal pools of the future agonists of the voluntary movement, occurring long in advance of the onset of the movement during preparation for its performance. Accordingly, we have termed the first phase of the latent period, during which this phenomenon occurs, the phase of the "pretuning" increase in the reflex excitability of the motor pool.

To determine the possible role of the pyramidal system in the origin of this "pretuning" increase, we performed studies with 5 patients with a clinical picture of central pyramidal paralyses. In 2 of the cases, the patients were suffering from hemiparesis (one of the left side, the other of the right side) of the pyramidal type following an operation for the removal of a tumor (meningioma) from the sinistral and dextral frontal areas, respectively; in the other 3 cases, the patients were suffering from hemiplegia (2 of the right side, 1 of the left side) that developed as a result of injury. They all had difficulty performing voluntary movements of the hip and knee joints of the affected extremity, but were incapable of any voluntary movement in the ankle joint (the latter condition being the criterion for their selection as subjects). The patients were instructed to attempt to perform extension of the paralyzed foot in response to a light stimulus. A test H-reflex was evoked at various delays after the movement signal was presented--0, 100, 150, 200, 250, or 300 msec. The test H-response was evoked 5-10 times for each delay. In 3 cases, the test H-response was evoked only with delays of 200, 250, and 300 msec. The average results of these experiments are given in Table 2.

As the data in Table 2 indicate, during the entire latent period of the "imaginary" movement of the paralyzed foot, there are no noticeable changes in the amplitude of the test H-response, which does not differ from the resting control response (cf. H-responses at rest and at various intervals following movement signal). The patients clearly attempted to perform the movement requested, and during their efforts, slight movements were observed in the hip and, at times, the knee joint. At the same time, the average amplitude of the test H-response in the paralyzed extremity evoked with zero delay after presentation of the signal to move the healthy leg was significantly greater in all 5 patients than in conditions of rest.

Table 2. Average Amplitudes of Test H-Resonse at Various Moments During Latent Period of "Imaginary" Movement by Patients with Pyramidal Paralysis of Foot (in % of Maximal Resting Amplitude)

Patient	Control, resting	Delay from movement signal to application of test stimulus (msec)					
		0	100	150	200	250	300
K.	37.7	36.6	39.9	40.5	35.2	33.3	38.4
	43.9	--	--	--	42.7	39.7	44.6
I.	30.0	35.5	32.2	36.6	40.2	30.6	29.6
	33.3	--	--	--	30.2	28.8	36.5
T.	40.1	37.8	43.3	44.5	39.9	42.2	47.6
	34.4	--	--	--	30.6	37.6	32.2
K.	26.6	28.8	33.6	25.9	29.5	30.0	27.4
S.	43.3	39.5	34.9	36.9	40.2	38.6	37.9

Thus, in patients with pyramidal lesions, the "ready-to-move" state is manifested prior to movement of the healthy limb in a background increase in the reflex excitability of the GM motoneurons of the paralyzed foot. However, an attempt to perform a movement of the paralyzed foot is not accompanied by a "pretuning" increase in the reflex excitability of these motoneurons. We may therefore conclude that different supraspinal structures (mechanisms) are apparently responsible for the "pretuning" and "background" increase in the reflex excitability of the motor nucleus.

The "Tuning" Phase

If less than 60 msec remain before the onset of voluntary extension of the foot, the amplitude of the H-response of the GM (the future agonist) rises above its amplitude in the first "pretuning" phase of the latent period, the increase being proportional to the proximity of the test H-response to the onset of the movement (see Fig. 11). The gradual increase in motoneuronal reflex excitability during the last 60 msec of the latent period appears to be specific, since it is strictly confined to the onset of voluntary impulse activity and is observed only in the motoneuronal pools of the future agonists, e.g., in the GM pool only prior to foot extension, and in the ATM pool only prior to foot

Table 3. Average Amplitude of H-Response During Various Intervals Prior to Onset of Voluntary Movement (in % of Maximum Amplitude

Index	Interval from H-response to onset of EMG movement (msec)					
	100-95	90-85	80-75	70-65	60-55	50-45
X	43.4	44.0	43.0	42.5	55.1	68.3
σ	±7.1	±7.4	±7.6	±7.3	±7.1	±7.9

flexion (see Chapter 4, Section 2). Changes in the amplitude of the H-response of the agonist during the second phase of the latent period appear to stem from a "tuning" of motoneuronal excitability that is supraspinal in origin. We have thus termed the second phase of the latent period, beginning 60 msec prior to the onset of movement, the "tuning" phase.

It can be seen from the recordings in Figs. 11A and B and the graph in Fig. 11C that the amplitude of the test H-response increases quite uniformly during the second ("tuning") phase of the latent period, the steepness of the climb being inversely proportional to the initial control amplitude of the H-response. The greatest amplitude is recorded at the onset of voluntary agonist activity. If H-responses with an initial resting amplitude greater than 20% of the maximal are used as the test responses, their amplitudes during voluntary movement may exceed the amplitudes of the maximal H-response during rest. Special experiments were performed to determine precisely the duration of the "tuning" phase and its dependence on the overall length of the latent period of voluntary movement (Table 3).

As indicated by data averaged from 26 experiments (20 subjects), during intervals of more than 60 msec before voluntary movement, the average amplitude of the test H-response assumes a plateau value ("pretuning" phase). The first significant increase in the amplitude of the test H-response is observed 55-60 msec prior to voluntary movement. As mentioned earlier, as the onset of voluntary agonist activity draws closer than this, the amplitude of its H-response continues to increase. Thus, the duration of the supraspinal "tuning" of the reflex excitability of the spinal motoneurons of the future agonist is 55-60 msec (for convenience, we shall henceforth assume a duration of 60 msec). It is significant that the duration of the supraspinal "tuning" is not related to the total duration of the latent period, a conclusion that is borne out by the following facts.

First, the duration of supraspinal "tuning" is identical in subjects who exhibit latent periods of different lengths. For example, the average length of the latent period for 20 subjects was found to be 125-140 msec for 5 subjects, 140-160 msec for 9 subjects, 160-180 msec for 4 subjects, and longer than 180 msec for 2 subjects. Nevertheless, in all subjects, the first significant changes in the amplitude of the test H-response of the agonist were recorded 55-60 msec prior to the onset of voluntary movement.

Second, the duration of the "tuning" does not vary significantly during the repetition of movements by a subject, despite variations in the duration of the latent period (Mart'yanov, 1968). Using at least two different delays in each experiment between the movement signal and elicitation of the H-response, we were able to determine the duration of the supraspinal "tuning" for varying latent periods. Regardless of the delays used, the earliest significant increase in the test H-response was recorded 55-60 msec before the onset of voluntary movement. For example, in one experiment (subject V.), the test H-response was separated from the movement signal by 90 or 120 msec. In both cases, a significant rise in the H-response was observed 55-60 msec prior to the voluntary myogram. However, when a 90-msec delay was used, the increase in the H-response 55-60 msec prior to the voluntary EMG volley of the agonist occurred in movements with a latent period of 145-150 msec [90 + (55-60)], while with a 120-msec delay, the increase occurred in movements with a latency of 175-180 msec [120 + (55-60)], i.e., 30 msec longer than in the first case.

Third, the duration of the supraspinal "tuning" does not change as the latent period of voluntary movement is shortened as a result of training. It is known that daily, frequent repetition of a movement leads to a progressive shortening of the latent period during the first few days. According to our data (a study by S. Migacheva, graduate student at the State Central Institute of Physical Culture), the latent period was shortened by an average of 12.4% by the second day of experimentation (16 subjects). Despite this reduction, in every experiment the duration of the supraspinal "adjustment" remained constant (55-60 msec) on the second day.

In this series of experiments, supraspinal "tuning" was determined under conditions of transition from complete rest of the agonist motoneuron pools (absence of EMG) to voluntary impulse activity. It was important to determine whether such a "tuning" exists and, if so, what is its duration prior to a phasic movement performed after a static effort, i.e., when there is some voluntary impulse activity of the agonist motoneuronal pools prior to movement. In these experiments (Kots, 1969b), the subject pressed his foot against an inclined pedal, exerting a slight, constant effort that was monitored with the aid of a dynamometer. The recordings were made on a "Diza" myograph operating in the "special" mode (Fig. 12A). In this mode, the electrical activity of the MG was picked up by one electrode pair and recorded simultaneously on two channels: on the central channel, a continuous record was made by feeding photosensitive paper past the cathode beam (Fig. 12A, middle record, read from top to bottom); on the left channel, a transverse record was made of the beam sweep (Fig. 12A, left record, read left to right), which began at various intervals after presentation of the movement signal, depending on the delay chosen. The presentation of the light signal (movement signal) was indicated on the left edge of the photographic record by a vertical black line. The H-response of the Sol was recorded on the right channel, its amplitude being used as a measure of changes in the reflex excitability of the motoneurons of the agonist involved in the phasic movement (recorded on the left channel).

Both in the absence and in the presence of tonic impulse activity by the motoneurons of the future agonist, the testing stimulus has no effect on the time of appearance of the phasic EMG volley. This conclusion is based on two facts: First, in all experiments, the average latency of the phasic movement during H-reflex testing did not differ from the latency without testing. Second, the EMG volley of the phasic movement can commence at any interval after the test H-response (see Figs. 11A and 12A).

The records in Fig. 12A and the graph in Fib. 12B show that the basic results of this series of experiments are consistent with results described earlier. The latent period preceding the onset of a phasic movement performed after static effort can also be divided into two phases, depending on the change in the amplitude of the test H-response: the first, "pretuning" phase, which lasts until 60 msec before the onset of the phasic movement; and the second, "tuning" phase, which spans the last 60 msec of the latent period. During the first phase, the amplitude of the test H-response of the agonist is greater than under conditions in which the subject exerts only a static effort not preparatory to the performance of phasic extension (see Fig. 12A, traces 4 and 3). During the first, "pretuning," phase of the latent period, the amplitude of the test H-response experiences no noticeable changes (see Fig. 12A, traces 4-6, and the graph in Fig. 12B). During the last 55-60 msec preceding the onset of voluntary phasic movement, the amplitude of the test H-response increases in proportion to the proximity of the EMG volley of the phasic movement (see Fig. 12A, traces 7-11, and the graph in Fig. 12).

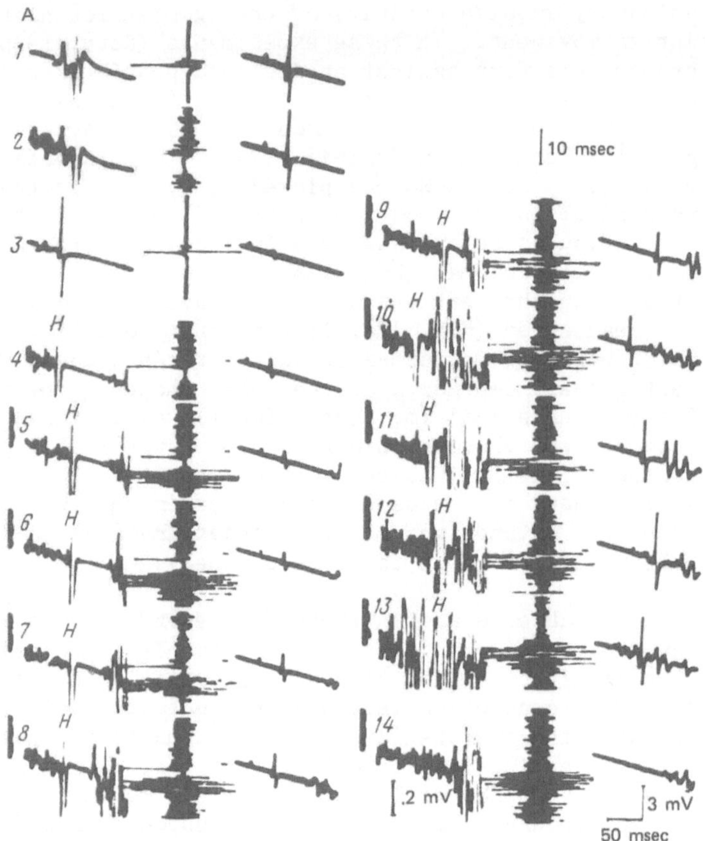

Fig. 12. H-response of Sol during latent period and at onset of voluntary extension of foot performed after voluntary static tensioning of GM. (A) Maximal H-response at rest (1) and during voluntary static muscle tension (2). Further explanation in text. (B) Change in test H-response of Sol during latent period of voluntary movement. Dots represent amplitudes of individual H-responses; circles represent average amplitudes.

The greatest amplitude of the Sol H-response is recorded during the voluntary phasic EMG volley (see traces 12 and 13).

Thus, the supraspinally governed transition from one level of voluntary impulse activity of the motoneuronal pool of the agonist

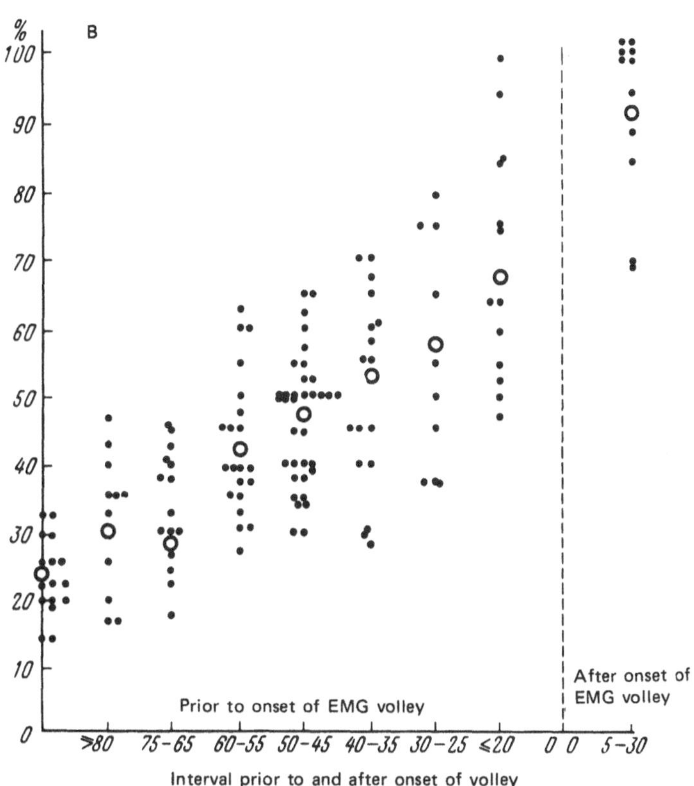

to a higher level is due to the preliminary "tuning" of the reflex excitability of the motoneurons of this pool, as is the transition from a state of complete rest to voluntary impulse activity. In both cases, the duration of the supraspinal "tuning" is identical and ranges from 55 to 60 msec.

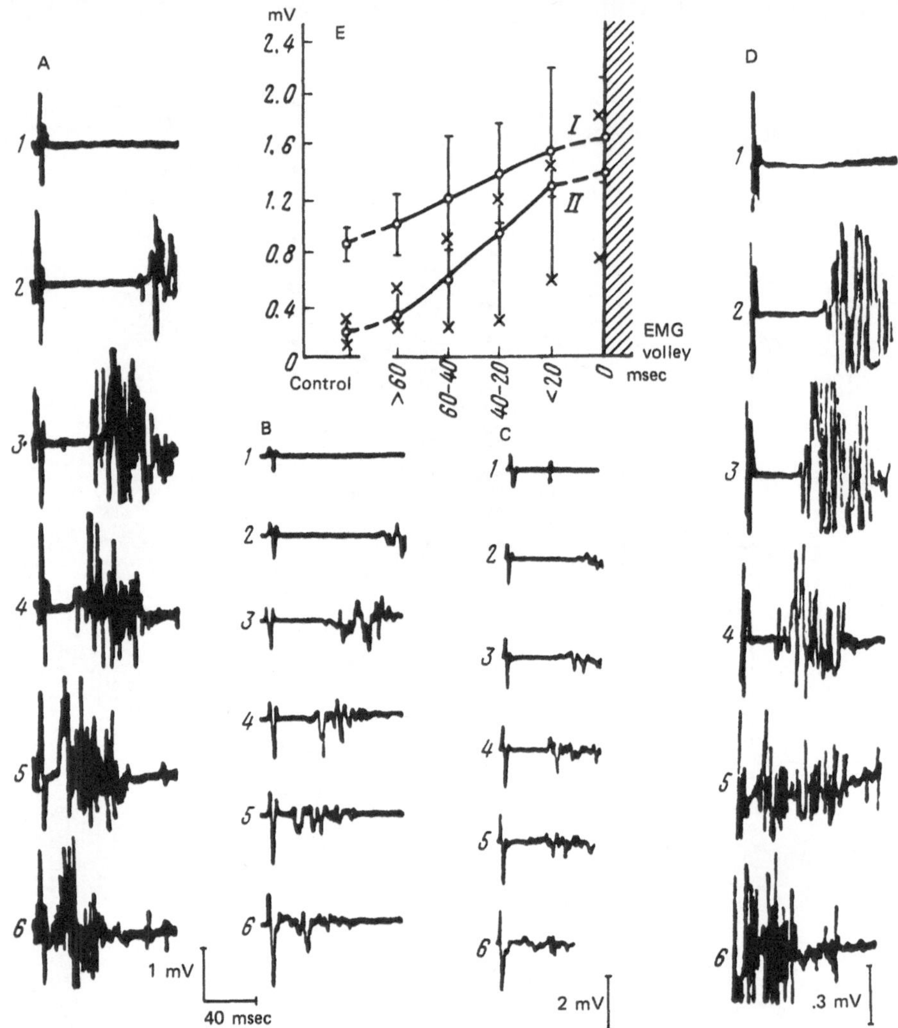

Fig. 13. Achilles tendon reflex during latent period of voluntary foot extension. (A-D) At rest (1); approaching onset of EMG volley of MG (2-5); at onset of voluntary EMG volley of MG (6). At start of beam recordings: reflex response. (A,D) EMG of MG; (B,C) EMG of Sol. Traces with the same numbers in A and B and in C and D were made simultaneously. (E) Graphic representation of average results.

Using the Achilles tendon reflex (T-reflex) as a test reflex, we found that the changes in this reflex during the latent period of voluntary movement have all the features characteristic of the H-reflex (Fig. 13): (1) the amplitude of the T-response is increased over its resting amplitude and exhibits no noticeable changes during the first ("pretuning") phase of the latent period; (2) during the last 50-70 msec of the latent period, the amplitude of the T-response increases in proportion to its proximity to the onset of voluntary movement ("tuning" phase); (3) the amplitude of the T-response increases quite uniformly during the "tuning" phase, the increase being inversely proportional to the initial amplitude of the test T-response; (4) the greatest amplitude of the T-response is recorded during the voluntary EMG volley: at a sufficient strength of the test stimulus, the amplitude of the T-response during voluntary movement is greater than the maximum amplitude of the T-response at rest.

Analogous results were obtained in a series of subsequent studies (Gurfinkel' and Pal'tsev, 1965; Naidel' and Pal'tsev, 1965). In several studies a comparison was made between the dynamics of the H- and T-responses prior to the onset of voluntary movement, although results were contradictory. For example, according to data obtained by Pierrot-Deseilligny et al. (1971), the H- and T-responses of the GM undergo parallel changes during the course of the latent period. According to the findings of Coquery and Coulmance (1971) in a test situation involving a choice, the T-reflex increases during the latent period, while the H-reflex does not. However, the latter study made use of a test H-reflex with an amplitude close to maximal, which precluded the possibility of appreciable changes in the reflex. It appears that further investigations in this area are required.

The "Triggering" Phase

By utilizing the H-response of the MG muscle as the test response and analyzing components I and II separately (see Chapter 2, Section 4), we discovered differences in the dynamics of changes in the reflex excitability of fractions of "slow" and "fast" agonist motoneurons during the organization of voluntary movement (Kots and Krinskii, 1968). Figures 14A and B show fragments of an experiment in which four different H-responses of the MG were used as the test responses; records of two of these responses are given--from zones I and III. To monitor the two components of the H-response, the records were made simultaneously at different amplifications (a and b). The records are presented in chronological order both prior to and subsequent to the onset of the voluntary EMG volley. The results of the experiment are shown graphically in Figs. 14C and D. Curves Ia and IIa in Fig. 14C correspond to components I and II of the H-response of the MG, the records of which are shown in Fig. 14A; curves Ib and IIb, to

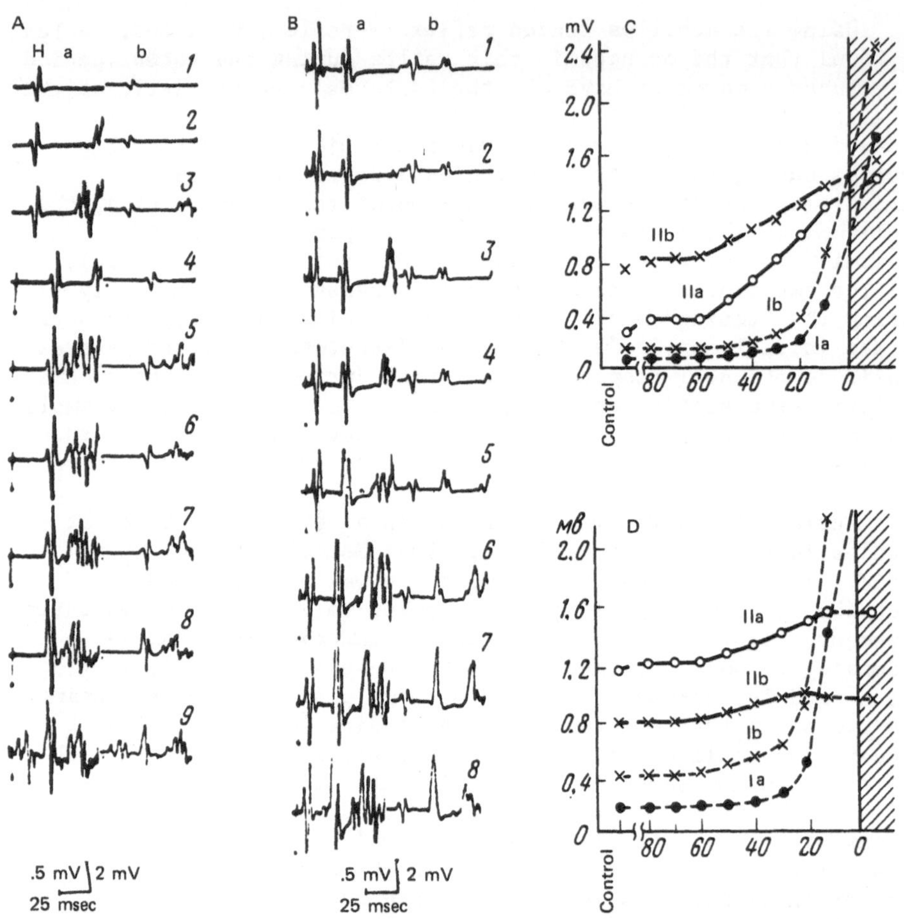

Fig. 14. H-response of MG during latent period and at onset of voluntary foot extension. Test stimulus 18 (A) and 28 V (B); (C,D) change in two components of H-response of MG during course of latent period. Explanation in text.

components I and II of the strong test H-response; curves Ia and IIa in Fig. 14D, to components I and II of a still stronger H-response of the MG; and curves Ib and IIb, to components I and II of the strongest H-response, the recordings of which are shown in Fig. 14B.

Figures 14A and B give the records (upper traces) of the test H-response of the MG during the latent period long in advance (>70 msec) of the onset of voluntary movement. Even in this early "pretuning" phase of the latent period, the amplitude of components I and II is greater than at rest. Like the H-response of the Sol, neither component of the MG H-response changes significantly during the first phase of the latent period. During the last 60 msec of the latent period (the "tuning" phase), there is a steady increase in the amplitude of both components as the onset of movement draws near. The changes in the amplitudes of components I and II are not parallel during this interval.

The amplitude of component II of the H-response of the MG (like that of the H-response of the Sol) increases continuously and quite uniformly during the last 60-msec interval of the latent period (see curves II in Figs. 14C and D).

Like the amplitude of the H-response of the Sol, the degree of increase in the amplitude of component II of the H-response of the MG during the last 60 msec of the latent period is inversely proportional to its initial (control) amplitude. This is shown graphically in Figs. 14C and D by the decrease in the slope of line II (cf. lines IIa and IIb in C and IIa and IIb in D).

The amplitude of component I of the MG H-response increases relatively slightly in the interval from 60 to 35 msec prior to the voluntary EMG volley and rises sharply during the last 25-30 msec of the latent period. Thus, the curve describing changes in the amplitude of component I of the MG H-response during the last 60 msec of the latent period consists of two parts, with a discontinuity approximately 30 msec prior to the onset of the voluntary EMG volley (see curves I in Figs. 14C and D).

The increase in component I of the H-response of the MG during the last 30 msec of the latent period and at the onset of movement is proportional to the magnitude of its initial amplitude. For example, if the amplitude of component I at rest equals approximately 15% of the maximal, it climbs to about 30% of the maximal during the "pretuning" phase of the latent period (see Fig. 14A, trace 1). Despite the significant increase, particularly during the last 30 msec of the latent period, the amplitude of this component prior to movement does not exceed the maximal resting value (traces 2-7). At the onset of voluntary EMG volley, it is approximately 25% greater than the maximal resting value (traces 8 and 9). In the test MG H-response with a resting amplitude of component I of about 80% of the maximal (Fig. 14B), this component begins to exhibit a significant increase only during the last 30 msec of the latent period (traces 5-7). Over this interval, the amplitude of component I increases such that just prior to the onset of the EMG volley (or during the onset), it has a value of 400% of the maximal resting value (trace 7b). During the voluntary EMG volley, the amplitude of component I may exceed the maximal resting amplitude by a factor of 5-7 (trace 8b) and equal up to 90% (or even 100% in rare cases) the maximal amplitude of component I of the M-response of the MG.

It must be noted that during the period of significant growth of the amplitude of component I, the amplitude of component II of the same MG H-response either attains a plateau value or even declines somewhat (e.g., see traces 8 and 9 in Fig. 14A, traces 5-8 in Fig. 14B, and graphs IIa and IIb in Fig. 14D). There are two possible explanations for this phenomenon: (1) An increase in the number of reflexly discharging "fast" motoneurons (increase in amplitude of component I of MG H-response) leads to a strengthening of the antidromic inhibition (through the Renshaw cells) of the "slow" motoneurons (see Chapter 2, Section 4). (2) The phenomenon results from the algebraic summation of the partially superimposed electrical responses from asynchronously firing "fast" and "slow" motor units. If the first hypothesis is correct, the decrease in component II of the H-response of the MG ought to be accompanied by a simultaneous decline of the H-response of the Sol, since both these responses are associated with the reflex discharge of "slow" motoneurons that are effectively inhibited through the recurrent collaterals of the "fast" motoneurons (Granit et al., 1957a; Kuno, 1959). However, the decline of component II of the MG H-response during a strong increase in component I is not accompanied by any decrease in the H-response of the Sol. It is most probable, therefore, that the decrease in component II under such conditions is purely a peripheral result of the partial superposition of two electrical responses with their algebraic sum in the "overlap zone."

The contrast in changes in the reflex excitability of "slow" and "fast" motoneurons of the GM apparently reflects two types of influence to which the motoneuronal pool of the future agonist is subjected. One of these influences is seen in the "tuning" increase in the reflex excitability of future-agonist motoneurons, which precedes their impulse activity by approximately 60 msec and has a comparable effect on the "slow" and "fast" motoneurons of the agonist. The reflex excitability of both species of motoneuron begins to change simultaneously, increasing uniformly as the onset of their impulse activity draws near.

The last 25- to 30-msec interval of the latent period displays the effects of another type of influence, one that is manifested in a sharp increase in the reflex excitability of the "fast" motoneurons, and that probably is superimposed on the uniform "tuning" increase in their reflex excitability. At the same time, the effects of this type of influence are not noticeably reflected in the course of the smooth "tuning" increase in the reflex excitability of the "slow" motoneurons. These later effects of the action of the second type of influence on the motoneuronal pool of the agonist are confined to the onset of its impulse activity, and may thus be designated as "triggering" influences. This presents two possibilities. Are these "triggering" influences on the agonist motoneuronal pool the result of a special type of supraspinal influence that is distinct from the supraspinal "tuning" influences, i.e., possess a different supraspinal origin, their own descending pathways, etc.; or does a single supraspinal source (or sources) activate two different (e.g., spinal) mechanisms that act separately (with a time delay) on the spinal motoneurons? One way of answering this question is to study the differences in the effects of different supraspinal influences on the "fast" and "slow" motoneurons.

Pyramidal and Extrapyramidal Influences on the "Fast" and "Slow" Motoneurons

Henneman et al. (1965b) found that the recruitment of motoneurons usually obeys the "size principle" during a variety of spinal reflexes. The size principle also operates for stimulation of supraspinal structures (Somjen et al., 1965), regardless whether the stimulation is ipsi- or contralateral, whether it induces mono- or polysynaptic activation of flexoral or extensoral motoneurons, or whether the stimulus is applied to the brain stem, cerebellum, motor cortex, or subcortical nuclei (striatum body and globus pallidus). Of two motoneurons the activities of which are recorded simultaneously from the ventral root filament, the smaller motoneuron (with a smaller-amplitude axon spike) responds at lower stimulus intensities than the larger motoneuron. This fact led Henneman and his associates to conclude that in the motoneuronal pool, the properties of the motoneurons themselves (their size) determine the order of their activation and inhibition. These authors hypothesized that the input from any afferent (for a given motoneuronal pool) system is uniformly distributed among all motoneurons of the pool, such that each motoneuron receives an equal share of the total input of any afferent system acting on the pool. Henneman and co-workers point out that in such a case, the effectiveness of synaptic bombardment must be different for motoneurons of different sizes due to variations in the input resistances of their membranes (Shapovalov, 1966; Burke, 1968).

At the same time, there are experimental data that show that in the cat (Sasaki and Tanaka, 1964; Grillner et al., 1970) as well as in the monkey (Preston and Whitlock, 1963; Clough et al., 1968), activation through different pathways evokes different effects in the "fast" and "slow" motoneurons of the MG and Sol muscles, and that the difference is not always consistent with expectations based on Henneman's "size principle." Activation of the stretch receptors (or the afferent pathways emanating from them) evokes a much greater facilitation of the "slow" motoneurons of the Sol than of the "fast" motoneurons of the MG (see Chapter 2, Section 4); stimulation of extrapyramidal structures (in particular the vestibular nuclei) leads to the more or less equal facilitation of both types of motoneurons (Sasaki and Tanaka, 1964; Grillner et al., 1970), while activation through the pyramidal tract evokes considerable facilitation of the "fast" motoneurons of the MG with no facilitatory effect on the "slow" motoneurons of the Sol (Preston and Whitlock, 1963; Sasaki and Tanaka, 1964). These findings led us to compare the effects of vestibulospinal (extrapyramidal) and voluntary influences on the "fast" and "slow" motoneurons of the GM in man.

In our investigations, the effects of vestibulospinal influences on GM motoneurons were studied by the method of galvanic stimulation of the vestibular apparatus, which was used as the preliminary conditioning stimulation. The H-reflex method was used to test changes in the state of spinal motoneurons of the Sol and MG in response to such a stimulus. The test H-reflex was evoked 100 msec after the onset of galvanic vestibular stimulation (for details, see Chapter 5, Section 1). The control ("pure") and conditioning ("vestibular") H-reflexes were alternated throughout the experiment.

The records from one of the experiments are given in Figs. 15A and B. Each line consists of four traces of H-responses elicited by equal stimulus strengths: the first and second traces from the left (below A) represent the H-responses of the MG; the third and fourth traces (below B) show the H-responses of the Sol; the middle (second and third) traces show the control ("pure") H-responses (maximal for a given stimulus strength); the outer (first and fourth) traces give the vestibular H-responses (also maximal for a given strength of the test stimulus). The records in the figure are arranged from top to bottom in order of increasing strength of the electrical test stimulus: for the H-response of the Sol and component II of the H-response of the MG, lines 1-3 show the suprathreshold control responses; line 4, the submaximal; line 5, the maximal; and lines 6-9, the supramaximal responses; for component I of the H-response of the MG, lines 1-5 show the suprathreshold responses; line 6, the submaximal; line 7, the maximal; and lines 8 and 9, the supramaximal responses. The results of this experiment are shown graphically in Figs. 15C and D. As the results of this and similar experiments indicate, galvanic stimulation of the vestibular apparatus induces a more or less equal increase in the reflex excitability of the "slow" motoneurons of the Sol and MG and the "fast" motoneurons of the MG.

Thus, our data are in agreement with the findings of Sasaki and Tanaka (1964) and Grillner et al. (1970), who employed intracellular recording and monosynaptic testing in cats to determine that vestibular stimulation (of the Deiters' nucleus) leads to the almost equal facilitation of "slow" (tonic) and "fast" (phasic) motoneurons in the Sol and MG muscles.

On the other hand, as was mentioned earlier, during the organization and at the onset of voluntary movement, the reflex excitability of a fraction of the "fast" and "slow" motoneurons undergoes

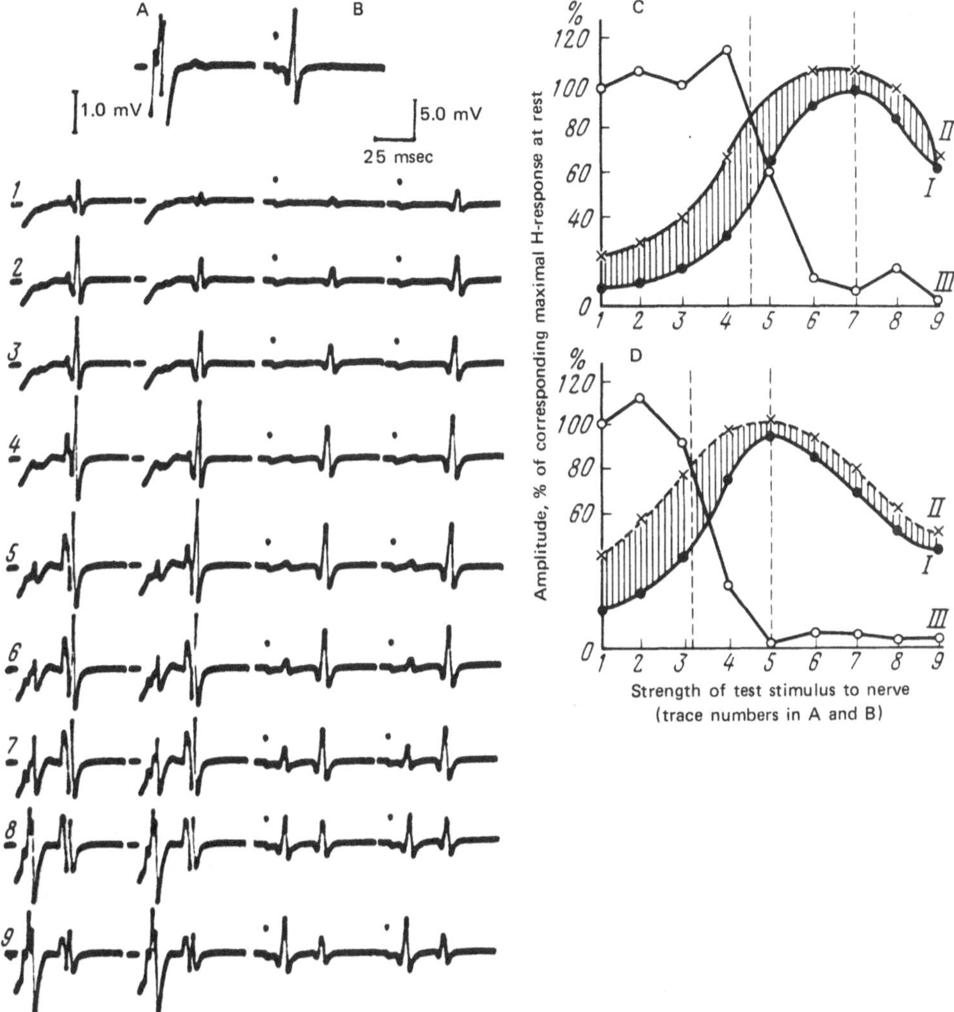

Fig. 15. Effect of electrical stimulation of the vestibular apparatus on the test H-responses of the MG (A,C) and Sol (B,D). Two upper traces under A and B: maximal M-response of MG and Sol, respectively. (C,D) Graphic representation of experimental results, corresponding to records given below A and B; (C) amplitude of component I of H-response of MG; (D) amplitude of H-response of Sol: (I) Amplitude of control responses; (II) amplitude of vestibular H-responses; (III) "vestibular addition." The vertical dashed lines correspond to 50% (left) and 100% (right) of the amplitude of the control H-response. Further explanation in text.

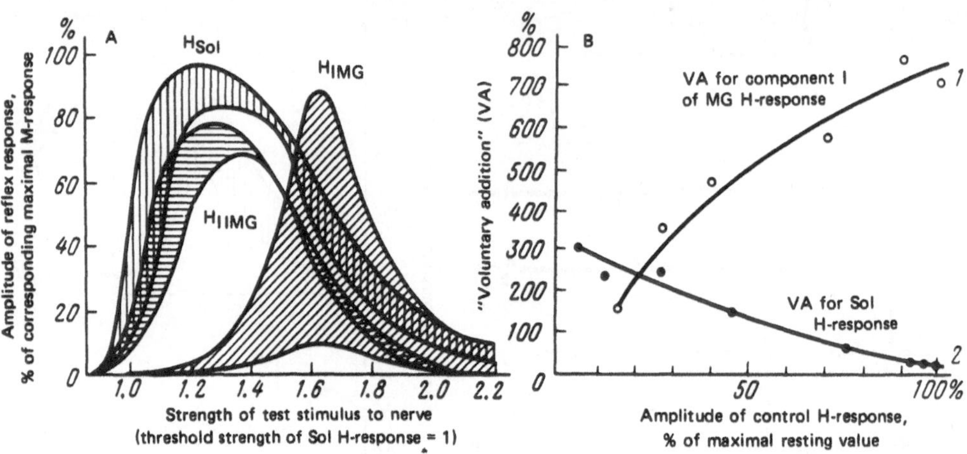

Fig. 16. Voluntary facilitation of the "fast" and "slow" motoneurons of the MG and Sol muscles.

different changes, the relative increase in the reflex excitability of the "fast" motoneurons (component I and H-response of MG) being much greater than that observed in the "slow" motoneurons (H-response of Sol and component II of H-response of MG). The graphic correlation of the results of one of the experiments shown in Fig. 16A provides a quantitative impression of the magnitude of the effect of voluntary facilitation for different types of motoneurons of the Sol and MG in man. In the graph, each set of two lines is joined by crosshatching--the lower line representing the amplitude of the test H-response at rest, and the upper line corresponding to the amplitude of the H-responses of the agonists during the first 60 msec of their EMG volley during voluntary extension of the foot. The vertical hatching corresponds to the H-response of the Sol, the horizontal hatching to component II of the H-response of the MG, and the diagonal hatching to component I of the H-response of the MG. As a comparison of the sizes of the hatched areas indicates, the relative increase in reflex excitability as a result of voluntary facilitation is significantly greater in the "fast" motoneurons of the MG than in the "slow" motoneurons of the Sol and MG.

These relationships are also illustrated in Fig. 16B, in which the ordinate represents the magnitude of the "voluntary addition," i.e., the percentage ratio of the difference in the amplitude of the H-response during voluntary movement and during rest to the amplitude of the H-response during rest. As the graphs show (curve 2), the magnitude of the "voluntary addition" is inversely proportional to the amplitude of the H-response of the Sol at rest: the smaller the amplitude of the control H-response, the greater its increase as a result of voluntary facilitation. There is a direct proportionality for component I of the H-response of the MG (curve 1): the greater the amplitude of component I at rest, the greater its increase during voluntary movement.

Such a relationship seems entirely plausible for the H-response of the Sol. Even during rest, most of the motoneurons of the Sol can be activated by an afferent test stimulus; therefore, as the latter is increased, the number of reflexly excited Sol motoneurons also increases (see Chapter 2, Section 1), and the possibility of "voluntary addition" is reduced. By contrast, the percentage of "fast" motoneurons in the MG that can be reflexly activated during rest is relatively small (see Chapter 2, Section 4). As experimental results show, during voluntary facilitation there is a sharp rise in the reflex excitability of these motoneurons. That the greatest increase in the amplitude of component I of the MG H-response is observed when a relatively high-intensity test stimulus is applied appears to provide additional confirmation of our hypothesis that the reflex excitation of "fast" motoneurons requires the activation of relatively high-threshold afferent fibers (see Chapter 2, Section 4). It is clear that during the application of weaker test stimuli that do not activate this fraction of the afferent fibers, the state of only a small portion of the "fast" motoneurons can be tested.

An investigation of the reflex excitability of "fast" and "slow" motoneurons in patients with pathological central lesions of the pyramidal system has furnished evidence that the difference in the level of voluntary facilitation of "fast" and "slow" motoneurons of the GM may be a result of differences in the degree of the effectiveness of pyramidal actions on these two motoneuronal species. In studies conducted by our colleague Makarova (1973), it was found that the reflex excitabilities of the motoneuronal pools of the Sol and MG muscles are altered to a different degree by central pyramidal lesions. In every case in which the two components of the H-response of the MG could be clearly distinguished in the healthy and affected extremities, it was determined that the amplitude (absolute and relative) of component I of the H-response of the MG was considerably greater on the paretic side than on the "healthy" side, while the asymmetry of the amplitudes of component II of the MG H-response is less pronounced, particularly in patients in whom the focus of the lesion is in the left hemisphere. On the average, the amplitude of the H-response of the MG in patients with pyramidal lesions is significantly higher in the affected leg than in the "healthy" leg or than in healthy subjects (Table 4). At the same time, the amplitude of the H-response of the Sol does not statistically exceed the corresponding values in healthy subjects: in patients with a sinistral focus, the average amplitudes of the H-responses of the Sol are equal in the healthy and affected

Table 4. Average Values of H_{max}/M_{max} (in %) of Sol and MG in Healthy Subjects and in Patients with Spastic Hemiparesis

Group	Number of examinees	Muscle	Side	H_{max}/M_{max}	σ	Significance of differences (Student's criterion)
Healthy subjects, 36–67 years of age	17	Sol	Right	61.4	±12.8	Insignificant
			Left	57.5	±16.4	
		MG	Right	23.5	± 7.0	Insignificant
			Left	19.8	± 7.9	
Patients with dextral focus of lesion	10	Sol	Healthy (right)	47.8	±10.3	Significant (P<0.001)
			Affected (left)	68.2	± 7.4	
		MG	Healthy (right)	25.0	± 7.6	Significant (P<0.001)
			Affected (left)	41.0	± 7.2	
Patients with sinistral focus of lesion	17	Sol	Healthy (left)	58.7	± 8.7	Insignificant
			Affected (right)	60.0	± 6.1	
		MG	Healthy (left)	26.6	± 6.7	Significant (P<0.01)
			Affected (right)	37.2	± 8.4	

extremities; the asymmetry of the amplitudes of the Sol H-responses in patients with a dextral focus is associated with a relative decrease in the amplitude of the Sol H-response in the healthy extremity.

Thus, central lesions of the human pyramidal system produce a substantial increase in the reflex excitability of "fast" motoneurons of the MG of the paretic extremity without appreciably changing the excitability of "slow" motoneurons of the Sol and MG. In this respect, clinical data are in agreement with experimental data, indicating that pyramidal influences have a more pronounced effect on "fast" motoneurons than on "slow" motoneurons.

The results obtained provide strong evidence that the two types of motoneurons that comprise the motoneuronal pool of the GM in man and other mammals have facilitatory connections with different reflex (spinal) and supraspinal afferent origins that differ in their effectiveness. These differences cannot be explained in terms of Henneman's "size principle," and force us to assume that at least some neural pathways have a nonuniform distribution of facilitatory inputs among the "slow" and "fast" motoneurons of the GM.

It should be noted, however, that Henneman did not have an adequate experimental basis for his hypothesis regarding the uniform distribution of afferent inputs among all motoneurons, because in his experiments he did not identify the motoneurons as "slow" or "fast," nor did he make an excitability comparison for the same pair (or several pairs) of concurrently recorded motoneurons for different modes of activation. Moreover, both our data and those in the literature indicate that populations of motoneurons that are activated primarily through different afferent inputs may not overlap to a significant degree. Therefore, the data obtained by Henneman et al. (1965a,b) justify only the following conclusion: among a population of motoneurons activated primarily through a given afferent input, the sequence of their activation is determined by their size. In other words, Henneman's "size principle" is apparently operative only within the limits of each such population (of each afferent input). It would be interesting to determine which portion (if any) of the motoneuronal size distribution curve Henneman's rule is valid in.

Given the fact that of the various supraspinal descending systems, the pyramidal pathways have a selective facilitatory effect on the "fast" motoneurons of the MG (Preston and Whitlock, 1963; Sasaki and Tanaka, 1964), we may assume that the "triggering" increase in the reflex excitability of the fraction of "fast" motoneurons of the GM agonist prior to voluntary movement is the result of pyramidal influences.

Judging by our results, the phenomena of "tuning" and "triggering" are abolished by clinical lesions of the pyramidal system. At the same time, these data do not necessarily indicate that spinal restructuring during the organization of voluntary movement is a direct result of pyramidal influences. It is quite likely, for example, that extrapyramidal mechanisms are also activated via the corticospinal system, and that these mechanisms determine the earlier "tuning" changes in the reflex excitability of the motoneurons of the future agonist of the voluntary movement. On these "tuning" influences are superposed the direct, "triggering" pyramidal influences that appear later. It is quite evident that both phenomena are abolished by the presence of central lesions of the corticospinal system. Another hypothesis is also possible: the two types of influences ("tuning" and "triggering") are produced by the action of the two pyramidal subsystems--the slow ("tonic") and fast ("phasic") subsystems--the presence of which has been convincingly demonstrated in cats in a large number of studies (Vasilenko and Kostyuk, 1965, 1966; Kostyuk and Vasilenko, 1968; Kostyuk, 1973).

3. THE SPINAL GAMMA LOOP

Analysis of the mechanisms responsible for the increase in the reflex excitability of the agonist motoneuron pool prior to voluntary movement has given rise to a hypothesis regarding the possible participation of the gamma loop in this phenomenon. Such a hypothesis is supported by a large volume of experimental data.

A number of studies (for surveys, see Granit, 1955, 1970; Matthews, P. B. C., 1964) have demonstrated the descending control of the state of fusimotor (gamma-) motoneurons by a variety of cerebral structures and all basic descending motor systems. As with the alpha-motoneurons, monosynaptic connections are established between the gamma-motoneurons and the pyramidal (Corazza et al., 1963; Wiesendanger and Laursen, 1966), rubrospinal (Appelberg and Molander, 1967), vestibulospinal (Pompeiano et al., 1967), and reticulospinal (Grillner et al., 1969) tracts. According to data obtained by Mortimer and Akert (1961), in monkeys, the cortical representation map of fusimotor neurons coincides strikingly with that of alpha-motoneurons. The cortical representation areas of alpha- and gamma-motoneurons are of approximately equal dimensions in cats and monkeys.

It was demonstrated in Granit's school that the gamma-motoneurons fire earlier and at lower stimulus strengths than the alpha-motoneurons for various modes of spinal and supraspinal reflex activation. These facts gave rise to the hypothesis that the "indirect" activation of alpha-motoneurons may be possible through the gamma loop in addition to their "direct" activation: the excitation of gamma-motoneurons amplifies impulsation from the muscle spindles, which in turn has a facilitatory effect on the alpha-motoneurons (Granit, 1955). In support of this hypothesis, it was found that interrupting the gamma loop by transection of the dorsal roots in the decerebrate cat removes the activation of alpha-motoneurons in some reflexes (cervicotonic, labyrinthine, pinna) without noticeably affecting the activation of gamma-motoneurons. Correlating the results of various experiments, Granit wrote: "In many if not most natural contractions hitherto studied, the gamma loop was first started, the nuclear bag afferents then facilitated, and the appropriate alpha motor neurons and direct alpha activation came last or together with gamma activity" (Granit, 1955, p. 268). Based on his findings, Granit concluded that the gamma system is utilized as an "ignition mechanism" to initiate movement.

It was hypothesized in most of the theories of the 1950's and 1960's regarding the mechanism of voluntary motor control that a voluntary movement is initiated and sustained by the activation of fusimotor neurons (gamma-motoneurons), which, together with the muscle spindles and monosynaptic reflex arc, comprised a follow-up length servomechanism (Merton, 1953).

In support of Granit's data, it was found in a number of later studies that the threshold of gamma-motoneuronal activation is lower than that of alpha-motoneurons for a variety of supraspinal influences (Calma and Kidd, 1959; Mortimer and Akert, 1961; Laursen and Wiesendanger, 1966; Carli et al., 1967; Maksimova, 1971; Granit, 1970). In a long series of different experimental situations, it was observed that the activation of alpha-motoneurons is preceded by the activation of gamma motoneurons. In particular, while experimenting with cats immobilized by Flaxedil, Buchwald et al. (1961) discovered that during avoidance conditioning in response to a conditioned signal, impulsation in gamma-afferents increases 10-20 msec earlier than in alpha-afferents. Dorsal root sectioning removes the conditioned reflex activation of alpha-motoneurons (Buchwald et al., 1964).

One of the basic experimental arguments in favor of the Granit-Merton theory has centered on the cessation of alpha-motoneuronal reflex activity following dorsal root section. However, if the disruption of the gamma loop is accomplished by ventral root section, no change is observed in the excitability of the alpha-motoneurons, and they can still be activated by the pinna reflex (which is not the case after dorsal root deafferentation) (Granit, 1955). On the other hand, in experiments with cats immobilized by Flaxedil, Severin (1966) disrupted the gamma loop by transection of the ventral roots and showed that deafferentation removes alpha-motoneuronal excitation previously induced by mild asphyxia or a pain stimulus. Hence, the role of the gamma loop is probably different for different modes of reflex activation of the alpha-motoneurons. Moreover, it is known (for a survey, see Kots et al., 1966) that even complete limb deafferentation by transection of the dorsal roots does not eliminate the possibility of alpha-motoneuronal activation or the performance of purposive limb movements (Munk, 1909; Asratyan, 1953; Twitchell, 1954; Veber, 1964; Drozdova, 1964, 1966), including conditioned reflex movements (Yankovska and Gurska, 1960; Knapp et al., 1963; Ovsyannikov, 1967). It is significant that these movements are observed immediately after the operation, which precludes the possibility of training. These facts allow us to conclude that alpha-motoneurons may be activated without participation of the gamma loop during the performance of at least a certain class of movement.

As a matter of fact, it has been shown in a number of recent experiments with anesthetized and decerebrate preparations that during movements elicited by the electrical stimulation of central structures, the fusimotor (gamma) system is not always activated along with the alpha-motor system on evocation of simple reflex and more complex motor acts (for surveys, see Matthews, P. B. C., 1964; Granit, 1970).

It has been determined that supraspinal influences on alpha- and gamma-motoneurons need not be interdependent to a significant degree, and may in fact be accomplished through different channels. In particular, the results of Diete-Spiff et al. (1967) indicate that in the cat, different vestibular nuclei possess different connections with alpha- and gamma-motoneurons. Thus, stimulation of the Deiters' nucleus produces a parallel change in the activity of both types of motoneurons, while stimulation of the medial and descending vestibular nuclei (as well as of the VIIIth cranial nerve) significantly changes the activity of the gamma-motoneurons, but has little effect on the activity of the alpha-motoneurons. According to the findings of Poppele (1967), a head turn in the decerebrate cat with unilateral destruction of the labyrinth elicits largely independent alpha- and gamma-motoneuronal responses from the GM, suggesting their independent activation. The gamma-motoneuronal responses were recorded in the absence of an alpha-motoneuronal response, whereas the latter could be observed even after the dorsal roots were transected.

Kato et al. (1964) discovered opposite pyramidal effects in the alpha- and gamma-motoneurons. However, since these authors recorded activities from the ventral roots, they could pick up impulses from alpha- and gamma-motoneurons belonging to different muscles. In a study by Koeze et al. (1968), interesting facts are presented that indicate the possibility of independent motor cortex control of alpha- and gamma-motoneurons in baboons. These authors determined that weak, short, high-frequency stimulation of the motor cortex, which elicits brief alpha-motoneuronal activation primarily through the direct corticomotoneuronal system, has no effect on the muscle spindles. When the stimulation of the cortex is strengthened, their thresholds to cortical stimulation are of the same order of magnitude. The most compelling evidence for the independent activation of alpha- and gamma-motoneurons was obtained in experiments with continuous, relatively low-frequency cortical stimulation with the activity of spindle afferents recorded during the occurrence of clonic epileptiform bursts.

In recent years, researchers have differentiated two types of influences on muscle spindles; the two types are accomplished through two types of gamma-fiber--dynamic and static (for a survey, see Matthews, P. B. C., 1964). The dynamic gamma-fibers amplify the response of primary endings to phasic muscle stimulation, while the static fibers depress this response. At a constant muscle length, the static gamma-fibers accelerate the discharge of primary and secondary muscle-spindle endings. Noting the presence of two types of fusimotor neurons in the fusimotor system, a number of researchers renewed their studies on supraspinal influences on muscle spindles. It was found in these studies that the supraspinal centers of the cat have a differentiated control function with respect to the dynamic and static sensitivity of muscle-spindle sensory endings (Jansen and Matthews, 1962a,b; Appelberg, 1962; Appelberg and Emonet-Denand, 1965; Vedel, 1966; Appelberg and Molander, 1967; Grillner et al., 1969).

According to the findings of Vedel (1966), rhythmic stimulation of the motor cortex in cats increases the dynamic, but not the static, sensitivity of the stretch response of muscle-spindle primary endings. Pyramidotomy destroys this effect. On the basis of these findings, Vedel concluded that the feline pyramidal tract controls only the dynamic fusimotor neurons. However, it was demonstrated in the most recent study by Yokota and Voorhoeve (1969) that rhythmic stimulation of the cat motor cortex increases the static sensitivity of secondary endings, as well as the static and dynamic sensitivity of primary endings of the spindles of antigravity forelimb muscles. These effects are evoked from the same areas that, when stimulated, induce the contraction of muscles containing these muscle spindles. But the threshold of muscle-spindle activation is lower than that of alpha-motoneuron activation. These influences are mediated principally through the fastest pyramidal fibers. On the basis of their data, Yokota and Voorhoeve concluded that the motor cortex controls the activity of both static and dynamic fusimotor neurons via the pyramidal tract.

Appelberg and Molander (1967) observed a selective increase in the dynamic sensitivity of muscle-spindle primary endings on stimulation of the red nucleus or inferior olive. According to their data, the regulation of dynamic fusimotor neurons may be accomplished via the rubrospinal tract. During stimulation of lower brain stem structures--Deiters' nucleus or the medial longitudinal fasciculus leaving the reticular pontile formation--Grillner et al. (1969) observed the monosynaptic activation of alpha-motoneurons and the parallel excitation primarily of the static fusimotor motoneurons of the extensors.

Thus, data obtained in recent years indicate the complex structure of the supraspinal control of spinal alpha- and gamma-[fusimotor] motoneurons. Such control may be accomplished more or less independently for the alpha- and gamma-motoneurons, and may be differentiated in terms of the two types of fusimotor motoneuron.

One goal in our investigations was to determine the role of supraspinal gamma activation in the organization of a phasic voluntary movement in man or, more precisely, to determine whether the participation of the gamma loop is essential in the initial voluntary activation of alpha-motoneurons. With this goal in mind, we studied the effects of the temporary disruption of the gamma loop by means of tourniquet-induced ischemia of the extremity. Using this technique, we examined the characteristics of the supraspinal "tuning" of agonist motoneurons and the duration of the latent period (Kots, 1969, 1970). It was shown earlier (see Chapter 2, Section 2) that tourniquet ischemia leads to the development of a conduction block in H-reflex afferent fibers earlier than in efferent motor fibers. This period of the preferential occlusion of afferent fibers of the gamma loop was utilized in studying the role of the gamma loop in the initiation of voluntary alpha activity.

To evoke the test H-reflex, the stimulating electrode was positioned in the popliteal fossa proximally to the tourniquet cuff (distal tourniquet, see Chapter 2, Section 2), which made it possible to test the state of motoneuronal reflex excitability during the preparation for voluntary movement against a background of ischemic deafferentation (disruption of the afferent limb of the gamma loop).

The results of one experiment dealing with the dynamics of the reflex excitability of GM motoneurons during the latent period of voluntary foot extension are shown graphically in Fig. 17A. The average amplitudes of the test H-responses at various intervals prior to the voluntary EMG volley under control conditions are shown as solid circles in the graph. Averaging was based on no fewer than 10 H-responses. A general "tuning" increase in the amplitude of the test H-response of the agonist can be observed during the last 60 msec of the latent period (see Section 2 above).

Since the amplitude of the control H-response gradually increases under conditions of distal ischemia (see Chapter 2, Section 2), in experiments involving ischemic deafferentation, it was necessary to reduce the strength of the test stimulus periodically. In the graph, different symbols are used to indicate the amplitudes of H-responses elicited by a constant stimulus strength at different stages of distal ischemia (x's, H-responses at 2nd to 5th minutes of ischemia; squares, at 6th to 11th minutes; triangles, at 12th to 16th minutes; open-circled x's, at 17th to 21st minutes; solid-circled x's, at 22nd to 25th minutes; solid circles, at 26th to 29th minutes). In Fig. 17A, the average data on amplitude fluctuations of the test H-response of the agonist during the latent period and under conditions of tourniquet ischemia are represented by open circles.

As these data indicate, under conditions of ischemic deafferentation, there are no changes in the character or duration of supraspinal "tuning" of the reflex excitability of agonist motoneurons. As under normal conditions, during the first phase of the latent period, the amplitude of the test H-response of the future agonist exhibits no significant changes, but increases steadily during the second phase (last 55-60 msec of the latent period) as the onset of the voluntary EMG volley draws near. The greatest amplitude of the test H-response is recorded during the voluntary EMG volley.

It is of particular significance that the character of the supraspinal "tuning" undergoes no noticeable changes during the late stages of distal ischemia (after the 20th minute from the start of the ischemic period), since by this time there is already occlusion of a significant number of low-threshold H-reflex afferents, and blockade of even the higher-threshold afferent H-reflex fibers and tactile receptors begins to appear (Kots, 1970; see also Chapter 2, Section 2). Findings indicate that the amplitude of the test H-response of the agonist increases steadily during the last 60 msec of the latent period at any stage of distal ischemia (see the positions of solid-circled x's and the solid circles in the graph in Fig. 17A).

The results of a special series of experiments designed to measure the latency from presentation of the movement signal to the onset of voluntary agonist EMG activity are summarized in Table 5. Each average value is the result of 10 trials. As the data show, during the ischemic period, there are no significant changes in the latency of voluntary movement. The same results were obtained in measurements of motor reaction time during experiments with both H-reflex testing and distal ischemia.

Thus, ischemic deafferentation does not disrupt the usual course of "tuning" facilitation and has no effect on the onset of voluntary impulse activity by spinal alpha-motoneurons for the type of voluntary phasic movement studied.

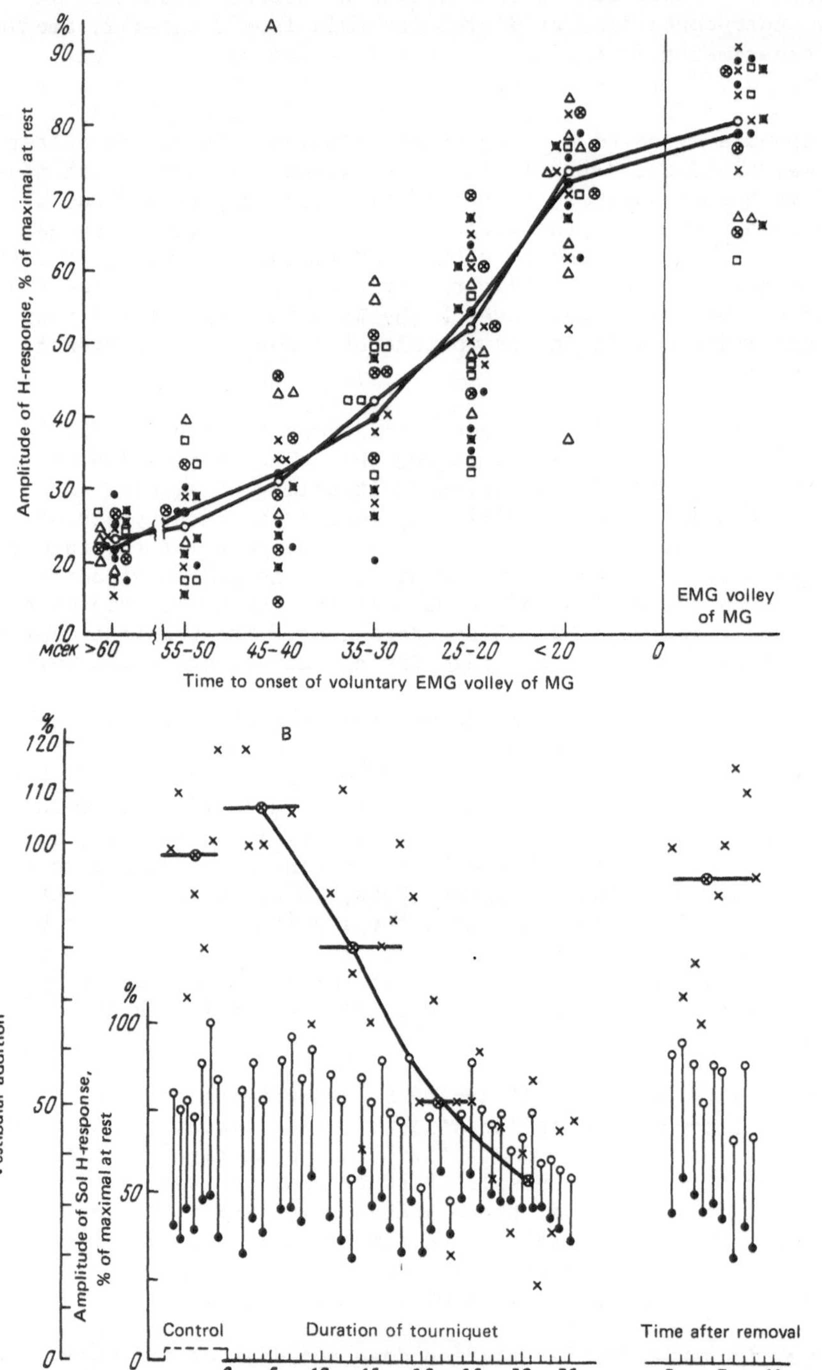
A
%
90
80
70
60
50
40
30
20
10
Amplitude of H-response, % of maximal at rest
мсек >60
55–50
45–40
35–30
25–20
< 20
0
EMG volley of MG
Time to onset of voluntary EMG volley of MG
B
%
120
110
100
50
0
"Vestibular addition"
Amplitude of Sol H-response, % of maximal at rest
Control
Duration of tourniquet
Time after removal
0
5
10
15
20
25
30
35
3
7
11
min

However, analysis of the results described above raised the following questions: To what extent is the afferent limb of the gamma loop disrupted by tourniquet ischemia? In other words, does ischemia produce a disruption of the afferent limb of the gamma loop that could alter the character of the supraspinal activation of alpha-motoneurons, given that this activation is dependent on the gamma loop; and can this change be determined by the methods employed? To resolve these questions, experiments were performed that were designed to reveal the influence of tourniquet ischemia on the magnitude of the vestibulospinal facilitatory effect evoked by galvanic stimulation of the vestibular apparatus.

Electrical stimulation (Andersson and Gernandt, 1956) as well as adequate stimulation of the vestibular apparatus by rotation (Totsuka et al., 1963) are known to significantly amplify the activity of gamma-motoneurons. The thresholds of activation in both cases are lower for gamma-motoneurons than for alpha-motoneurons.

Table 5. Latency (in msec) of Voluntary Movement During Tourniquet Ischemia

Subjects	Before tourniquet	During tourniquet (min) 3	10	17	23	27
M.	129±4.2	126±1.7	129±4.7	130±4.3	135±5.2	127±4.5
V.	157±6.7	149±6.5	139±6.1	148±7.7	146±6.7	141±6.4
Kh.	130±3.7	140±4.8	135±6.5	135±4.2	127±5.9	133±5.0
Z.	136±5.1	132±5.9	123±4.5	131±4.3	121±4.2	129±3.6
S.	147±6.7	133±6.6	159±4.1	153±3.4	159±7.2	154±7.6
M.	110±3.6	105±3.7	111±3.7	107±3.5	119±4.4	112±3.2
Average time	134.8	130.8	133.5	134.0	134.5	132.6

Fig. 17. Effect of ischemic "deafferentation" on voluntary (A) and vestibulospinal (B) facilitatory effects. (A) "Tuning" increase in reflex excitability of motoneuron pool of Sol during latent period and at onset of voluntary foot extension under control conditions (solid circles) and during tourniquet ischemia of limb (open circles).

On the basis of these findings, it could be assumed that the increase in the reflex excitability of human GM motoneurons evoked by electrical stimulation of the vestibular apparatus (see Chapter 5, Section 1) is also dependent to a certain degree on the activation of the gamma loop, and that therefore this phenomenon can be altered by ischemic disruption of the gamma loop.

For the 6th to 7th minute prior to ischemia, for the 30th to 35th minute of distal ischemia, and for 7 to 10 minutes subsequent to ischemia (usually beginning from the 2nd to 3rd minute after tourniquet removal), the control H-response and conditioned "vestibular" H-response were alternately evoked (0.1 sec after start of galvanic stimulation of ipsilateral vestibular apparatus) (see Chapter 5, Section 1).

The results of one of the experiments are shown graphically in Fig. 17B. Vertical lines connect the average values of the amplitudes of the "pure" or "control" H-response (lower, solid circles) and the "vestibular" H-response (upper, open circles) obtained over 1-min intervals. The x's correspond to the values of the "vestibular addition"* (VA) for a given minute, and the open-circled x's to the average VA for several minutes. The time period on which averaging of the VA is based is indicated by a horizontal line through the open-circled x's.

As the graph shows, 8-10 min after the onset of ischemia, the vestibulospinal facilitatory effect begins to decline and continues this decline throughout the ischemic period, such that 30 min after the onset of ischemia the VA is only about 30% of the VA prior to and during the first 5 min of ischemia (from 15 to 40% in different experiments). Following ischemia, the magnitude of the vestibulospinal facilitatory effect returns to its initial level.

The results of these experiments indicate that use of the ischemia and monosynaptic testing methods makes it possible to determine the consequences of disrupting the afferent limb of the gamma loop for the supraspinal facilitation of alpha-motoneurons. Hence, the results of experiments designed to test the latent period of voluntary movement under conditions of ischemic "deafferentation" permit us to conclude that the temporary ischemic disruption of the afferent limb of the gamma loop does not prevent the voluntary activation of spinal motoneurons. The participation of the gamma loop is apparently not essential for the "tuning" increase in the reflex excitability and initial activation of the spinal alpha-motoneurons of the agonists involved in the voluntary

*The "vestibular addition" is the difference, expressed in percent, between the amplitudes of the "vestibular" and "pure" H-responses relative to the amplitude of the "pure" H-response.

phasic movement. The gamma loop does not appear to play an essential role as the "ignition mechanism" for the initiation of phasic voluntary movement (Kots, 1969, 1970).

The same conclusion was reached by Valbo (1971), who recorded the impulse activity of single afferents of the median and tibial nerves in human subjects. On the basis of several criteria (reactions to muscle stretch, blow to tendon, etc.), he identified these afferents as muscle-spindle afferents. By determining the delay between the onset of EMG activity and the onset of spindle acceleration, Valbo found that during the performance of both slow-rising voluntary isometric contractions and of fast, short-lasting isometric contractions, acceleration of muscle-spindle discharge occurs simultaneously with or subsequent to the onset of voluntary muscular contraction (see also Burg et al., 1973). These findings led the author to reject the notion that the fusimotor system plays a leading role and to conclude that this system is activated during voluntary contractions in man, but does not participate in their initiation. According to Valbo, these contractions are initiated by descending impulses from supraspinal structures and their effects in the neuronal organization within the spinal cord.

Our findings and those of other researchers have led to the conclusion that voluntary movement is initiated by the primary activation of alpha-motoneurons. It is noteworthy that the most recent experimental findings have radically changed original concepts regarding the functional role of the gamma system in the control of movement. Fidone and Preston (1969) believe that the more effective functional connection between the motor cortex and the fusimotor neurons in comparison with the alpha-motoneurons, which was discovered by them in experiments with pyramidal cats, can compensate for the considerable delay attending the reflex activation of alpha-motoneurons via the gamma loop, i.e., can permit the spindle discharge to act on the alpha-motoneurons with greater coincidence with the arrival of longer-latency "direct" effects on these motoneurons. On the basis of this assumption, the authors conclude that "gamma leading" does not denote the initiation of movement through the gamma loop.

Many authors see the role of fusimotor activation in the prevention of any decrease in spindle discharge occurring during muscular contraction. This function is accomplished by the adjustment of gamma activity in accordance with the anticipated speed of muscular contraction (Matthews, P. B. C., 1964; Koeze, 1968; Koeze et al., 1968). The functional significance of the simultaneous coactivation of alpha- and gamma-motoneurons evidently includes the fact that it permits signalization from the muscle spindles concerning the length of the muscle and the rate of change during the performance of a movement to be relayed to the CNS, in particular to the sensory and the motor cortex (Phillips, 1966). Ascribing particular significance to the latter projections, Koeze et al. (1968) argue that the phylogenetic magnification of the role of the cortical motor control mechanism has been accompanied by a decline in the importance of cortical fusimotor activation in the functioning of the segmental servomechanism (the peripheral gamma loop), but an increase in the importance of this activation in the maintenance of feedback to the motor cortex: from the spindles through the cerebrallar cortex into the cortical motor control circuits (cortical gamma loop).

4. INTERNEURONAL (INHIBITORY) SYSTEMS OF THE AGONIST

As histological studies have shown, most of the connections of the corticospinal fibers in primates are comprised of their contacts with segmental interneurons of the base of the dorsal horn in the intermediate area. This region has expanded during the process of ontogenesis in monkeys, and from the chimpanzee (Kuypers, 1964) to the human (Schoen, 1964). Practically no physiological studies have been done on the descending connections with the interneuronal segmental apparatus in primates. Studies with lower mammals (usually cats) have made it possible to distinguish several segmental interneuronal systems that may be utilized in the cerebral control of agonist spinal motoneuronal pools:

-- Interneurons of the spinal reflex pathways, which are directly activated by the primary afferents and are connected (directly or via supplementary intraspinal synaptic relays) to the spinal motoneurons.

-- Interneurons of presynaptic inhibition, which participate in the depolarization of intraspinal primary-afferent terminals and of intra- and supraspinal fibers.

-- Fusimotor neurons (gamma-motoneurons), which regulate the activity level of the afferent limb (group I-a and II afferents) of a number of reflex pathways.

-- The Renshaw cells, interneurons that ensure the central (intraspinal) interaction of motoneurons.

All the basic descending motor systems can act on the interneuron groups listed above.

It was discovered as early as 1906 by Sherrington (1906) that the reflex effects in spinal preparations do not coincide with those in nonspinal animals. Sherrington assumed that spinal transection produces a depression of all reflexes below the point of transection. It was later discovered, however, that spinalization actually amplifies the flexion reflex and the reflex inhibitory effects, particularly the inhibition of the extensor patellar reflex (for a survey of early works, see Eccles, R. M., and Lundberg, 1959a). Fulton (1926) surmised that such a "release" is associated with the termination of inhibitory actions by supraspinal centers on the interneurons that mediate the spinal reflexes, a hypothesis that was later confirmed experimentally (Job, 1953; Eccles, R. M., and Lundberg, 1959a,b).

In particular, studies by Lundberg (1964, 1966) revealed that systems that accomplish supraspinal control by acting on interneurons associated with flexor reflex afferents are able to determine whether the volleys in these afferents will evoke inhibition or excitation in the motor nuclei of the flexors. Experiments with cats have shown that stimulation of the sensorimotor cortex via the pyramidal tract elicits the enhancement of polysynaptic reflex effects (excitatory and inhibitory) evoked by the stimulation of various primary afferent systems--I-a and I-b afferents, flexor reflex afferents (group II and III muscle afferents, skin and high-threshold joint afferents), skin afferents, and pressure afferents. The pyramidal facilitation of spinal reflexes in these experiments could not be explained by direct effects on the motoneurons; thus, it had to be assumed that this facilitation is associated with pyramidal effects on interneurons of the spinal reflex pathways. In fact, according to the findings of Lundberg et al. (1962), stimulation of the corticospinal tract in the cat produces an EPSP in all investigated interneurons of the spinal reflex pathways from the I-a and I-b afferents, flexor reflex afferents, and low-threshold skin afferents.

According to data obtained by Zadorozhnyi et al. (1970), the corticospinal influences on interneurons of the spinal reflex pathways do not have such a universal character as the influence discovered by Lundberg and his colleagues. According to Zadorozhnyi and co-workers, corticospinal influences in the cat are clearly expressed in interneurons activated by flexor reflex afferents, but they are considerably less pronounced or are completely absent in interneurons activated only by low-threshold muscle or skin afferents. Fetz (1968) also found differences in the pyramidal influences on different dorsal horn interneurons. Neurons of lamina IV and part of lamina V (Rexed), which are activated by natural tactile stimulation of the skin (Wall, 1967), are inhibited by pyramidal influences originating primarily in the postcruciate cortex. These inhibitory influences are partly presynaptic in nature. By contrast, pyramidal influences (primarily from the precruciate cortex) evoke direct facilitation of interneurons of lamina VI and a portion of lamina V that are activated by tactile stimulation of skin with wide receptive fields and by movements (Wall, 1967).

Besides the corticospinal (pyramidal) system, the interneurons of spinal reflex pathways can also be acted on by other descending motor pathways. In the cat, the rubrospinal system has an effect on interneurons of the segmental pathways that is largely analogous to the action of the corticospinal (pyramidal) system (Hongo et al., 1965, 1969; Kostyuk and Pilyavskii, 1969; Kostyuk, 1973).

The influence of the vestibulospinal tract on spinal interneurons of the reflex pathways was revealed in a study by Erulkar et al. (1966), who demonstrated the presence of spinal interneurons in the lumbar segments of the cat that react both to stimulation of the ipsilateral vestibular nerve and to stimulation of the dorsal root.

The reticulospinal pathways (dorsal reticulospinal tract) have a tonic inhibitory effect on the spinal polysynaptic reflex pathways (Holmquist and Lundberg, 1959; Engberg et al., 1965). On stimulation of the medial reticular formation, inhibitory effects are observed in the interneuronal apparatus of various spinal reflex pathways, and are particularly pronounced in interneurons activated by flexor reflex afferents (Kostyuk and Preobrazhenskii, 1967; Engberg et al., 1968).

The supraspinal centers can also act on the segmental interneurons of presynaptic inhibition through various descending motor pathways, evoking the depolarization of primary afferents and thereby controlling the level of afferent input to the segmental reflex pathways, ascending sensorimotor pathways, and relay nuclei (Lundberg, 1964, 1966; Wall, 1964; Eccles, J., 1966; Kostyuk, 1970, 1973). Depolarization of the primary afferents has been observed on stimulation of various supraspinal structures: the sensorimotor cortex (Carpenter et al., 1962a, 1963; Andersen et al., 1964), the red nucleus (Hongo et al., 1972), the cerebellum (Carpenter et al., 1966), and brain stem structures (Carpenter et al., 1962b, 1966; Erulkar et al., 1966; Cook et al., 1968). Stimulation of the sensorimotor cortex or the pyramids evokes the depolarization of particular primary afferent groups: depolarization is most pronounced in the flexor reflex afferents, is clearly observed in I-b muscle afferents and in low-threshold skin afferents, but is absent in I-a muscle afferents (Carpenter et al., 1962a, 1963; Andersen et al., 1964). Stimulation of the red nucleus in the cat (Hongo et al., 1972) produces a significant depolarization of the primary I-b, skin, and low-threshold joint afferents, and frequently a hyperpolarization of I-a afferents.

In experiments conducted by Erulkar et al. (1966), substantial dorsal-root potentials were recorded on stimulation of the ipsi- and contralateral vestibular nerves in cats. By the configuration of the response and the action of picrotoxin, they concluded that this phenomenon is related to the activity of those interneurons of the spinal cord that are activated by the vestibular system and are responsible for the presynaptic inhibition of spinal reflexes. The experiments of Cook et al. (1968) showed that stimulation of the VIIIth nerve does evoke the depolarization of the primary group I afferents from the flexors and extensors and of the large group II skin afferents, which is also associated with the presynaptic inhibition of reflexes mediated by these afferent pathways.

It was determined in a study by Carpenter et al. (1962b) that stimulation of the brain stem produces a dorsal root potential, and an investigation of the excitability of primary afferents has shown that the generation of this potential is accompanied by the depolarization of the I-a, I-b, and skin afferents. At the same time, the findings of these authors indicate that the descending systems emanating from the brain stem in the dorsal and ventral portions of the ventrolateral columns evoke the depolarization of the I-b and skin afferents, but not the I-a afferents. On stimulation of the reticular system of the brain stem, Carpenter et al. (1966) observed effects analogous to those described earlier by Erulkar and co-workers for stimulation of the vestibular nerve. Influences on the state of the primary afferents are transmitted via the ventral and dorsal reticulospinal tracts. Both the dorsal (Engberg et al., 1965) and ventral (Lundberg and Vyklicky, 1963) reticulospinal tracts act on the interneurons, exerting their influence on different primary afferent systems--I-a, I-b, and flexor reflex afferents.

It is hypothesized that presynaptic inhibition is more effective with respect to weak sensory influences, promoting the selection of more significant afferentation (Kostyuk, 1969). The supraspinal control of this mechanism apparently has the ability to activate or modulate (weaken or enhance) the mechanism in accordance with the motor problem at hand (see Chapter 1).

The presence of a supraspinal control of the state of the Renshaw cells was first demonstrated by Haase and Van der Muelen (1961). In experiments with decerebrate, deafferented cats, they established the possibility of the facilitatory and inhibitory control of the activity of Renshaw cells activated by antidromic stimulation, by the cerebellum and ventromedial reticular formation of the brain stem. By recording intracellularly from Renshaw cells in cats immobilized with gallamine, MacLean and Leffmen (1967) supplemented the findings of Haase and Van der Meulen by showing that antidromically evoked Renshaw cell discharge is inhibited by stimulation of the pericruciate cortex, ventral thalamus, and reticular formation of the midbrain and medulla oblongata. Transection of both pyramids removes the effect of cortical stimulation. In many cases, the Renshaw cells reproduce the discharge frequency of the supraspinal structures. The findings of these authors indicate that the most effective and long-lasting inhibition of the Renshaw cells is evoked by stimulation of the bulbar reticular formation; cortical inhibition is much shorter and less effective. MacLean and Leffman conclude that intensive interaction is possible between supraspinal structures and the Renshaw cells. The supraspinal control of the Renshaw cells can ensure both the diffuse activation of motoneurons and their fine modulation over various pathways.

According to data obtained by V. J. Wilson et al. (1964), both supraspinal and reflex influences (with a wide receptive field from all four limbs) evoke mainly inhibitory effects in the Renshaw cells. Impulses in the axon collaterals of the motoneurons have an excitatory effect on these cells. Analyzing the role of the Renshaw cells in reinforcing the contrast between the active motor pool and adjacent quiescent pools, V. J. Wilson (1966) perceives an analogy with lateral inhibition: recurrent inhibition is involved in the localization or "focusing" of the motoneuronal output. Supraspinal influences may amplify or weaken these effects of Renshaw cell action. The significance of supraspinal control with respect to the reciprocal interaction of antagonist muscles is examined in Chapter 4, Section 3.

Thus, all basic descending motor systems may have a facilitatory and an inhibitory influence on various spinal interneurons, evoking in them excitatory or inhibitory effects by direct transsynaptic action or by acting on the presynaptic pathways to the interneurons in question. Although a wealth of data has been gathered on the segmental neuronal mechanisms of cerebrospinal motor systems, almost all data have been obtained under conditions of the artificial stimulation of supraspinal structures or their descending pathways in immobilized animals. It is still not known, therefore, which of these mechanisms is utilized in a natural motor control situation. Answers to these questions can be obtained only by studies on the performance of natural movements.

The data in the literature support the hypothesis that the organization of voluntary movement includes, as one of its mechanisms, supraspinal actions on the segmental inhibitory interneurons connected to the motor nuclei of the future agonists. Our studies on the state of inhibitory interneurons during the organization of voluntary movement consisted of three series of experiments: (1) experiments with paired stimuli (double H-reflex), providing information on the state of spinal inhibitory interneurons activated via the afferent (orthodromic) pathway; (2) experiments involving supramaximal H-reflex testing; and (3) experiments designed to determine the duration of the "silent period" following an H-reflex evoked at the end of the latent period or at the onset of a voluntary movement. The latter two series of experiments make it possible to test (to a certain degree) the state of the spinal inhibitory system that is activated antidromically, i.e., the state of the Renshaw cell system.

<u>Inhibitory Interneuronal System of the Agonist That Is Activated by Orthodromic Stimulation</u>

It was determined in a study by Magladery and his collaborators (Magladery et al., 1951; Teasdall et al., 1952) that during paired stimulation of the TN, the amplitude of the second (test) H-response is decreased below the amplitude of the control H-response elicited by a single stimulus, the decrease depending on the interval between the first and second stimuli. This fundamental observation was confirmed repeatedly in later studies (Khazovskaya, 1965; for references, see Cook, 1968).

According to our findings, the degree of depression of the second H-response is dependent on the strength of the first and second stimuli: at a constant strength of the second (test) stimulus, the greater the strength of the first (conditioning) stimulus, the greater the decline in the amplitude of the second H-response. Depression of the second H-response is observed even after application of a conditioning stimulus that is subthreshold for elicitation of the H-response. In this respect, our findings are consistent with data obtained in experiments with cats (Brooks, C. M., et al., 1950) dealing with the existence of the phenomenon of "subsynaptic depression." The degree of depression of the second H-response is inversely proportional to its magnitude: for a constant conditioning stimulus strength, the greater the test H-response (the stronger the test stimulus), the smaller the degree of its depression.

When the interval between the two stimuli is increased, the amplitude of the second H-response does not return monotonically to the control value, as described by Magladery and his collaborators. After the strong depression phase during about 80-100 msec after the first stimulus, the amplitude of the second H-response begins to increase relatively quickly, at times reaching 50% of its initial amplitude approximately 150-250 msec after the first stimulus (Fig. 18A). But then the amplitude of the test H-response gradually declines once more--the second inhibition achieves a maximum about 400-600 msec after the conditioning stimulus. Then the test H-response begins to increase slowly and rather monotonically for 1 or 2 sec, at which time it has risen to about 70-80% of the control amplitude. Full recovery occurs after 15-20 sec. Ten seconds after the conditioning stimulus, however, the second H-response has already increased to about 95% of the control amplitude. In a paper published by McLeod (1969), we found data that are in complete agreement with ours. The curve relating

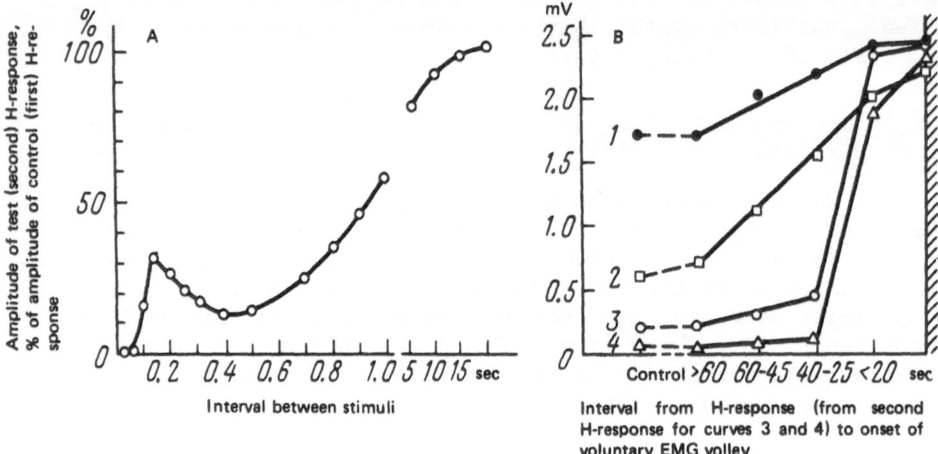

Fig. 18. Paired stimulus testing of agonist nerve during rest (A) and prior to voluntary movement (B). In A, the strengths of the 1st and 2nd stimuli are equivalent, and elicit an H-response from the Sol (resting) with an amplitude about 70% of the maximal.

changes in the amplitude of the test H-response to the intervals following the conditioning H-reflex is very similar to analogous curves for the monosynaptic reflex obtained in experiments with cats (for a survey, see Cook, 1968), thus providing additional evidence for the monosynaptic nature of the H-reflex (see Chapter 2).

The depression of the second H-response during paired stimulation may be the result of a number of factors: (1) an antidromic blockade in electrically excited motoneuronal axons occurring during interstimulus delays up to 20 msec and in the presence of an M-response to the first stimulus; (2) a depression of the excitability of motoneurons participating in the first reflex (refractoriness); (3) the influence of inhibitory interneurons activated by the first afferent stimulus; (4) the inhibitory influence of Renshaw cells activated by the first reflex motoneuronal volley or by the antidromic volley from the first stimulus (in the presence of an M-response to the first stimulus); (5) the presynaptic inhibition of afferent fibers of the H-reflex arc; (6) a decrease in facilitatory effects from the muscle spindles and an increase in inhibitory effects from the tendon receptors produced by the first muscular contraction.

The action of the first factor is observed (as mentioned) only during delays less than 20 msec, that of the sixth factor during delays longer than 30 msec. The second factor is not necessarily involved in the depression of the second H-response, since its participation is observed even when the first stimulus elicits no H-response, or one with an amplitude much smaller than the H-response evoked by a solitary second stimulus; i.e., the first stimulus elicits a response from a much smaller number of motoneurons than the second stimulus by itself. Thus, the activation of spinal interneurons with an inhibitory effect on the monosynaptic H-reflex plays a basic role in the depression of the second H-response during paired stimulation (factors 3-5 and, in part, factor 6). In studies done in humans, it is impossible to ascertain which of these inhibitory interneural systems determines the depression of the second reflex or to what degree; therefore, it may be accurate to apply a more general concept--a spinal inhibitory system the action of which can be associated with the phenomenon described. Inasmuch as the degree of depression of the second (test) H-response during paired stimulation depends on the state of the spinal inhibitory system, it may be possible to draw some conclusions on the dynamics of the state of this system during the organization of voluntary movement by the changes in the test H-reflex observed during the latent period (Kots, 1964).

Figure 18B gives a graphic representation of the results of one of the experiments involving paired stimulation of the agonist nerve. It can be seen that the amplitude of the single test H-response of the agonist increases quite uniformly throughout the last 60 msec of the latent period of voluntary movement (curves 1 and 2). During paired testing, the amplitude curve of the 2nd H-response appears to consist of two parts, with a sharp break in the interval 25-30 msec prior to the onset of voluntary movement. The line connecting the average amplitudes of the 2nd H-responses is much flatter during the first half of the last 60 msec of the latent period than during the second half of this interval (curves 3 and 4). A comparison of the amplitude curves of the solitary H-response and the 2nd H-response during paired stimulation reveals that the sharp increase in the 2nd H-response on paired stimulation during the last 25-30 msec of the latent period cannot be explained solely by a continuing increase in the excitability of the motoneurons of the future agonist. If this were the only factor, then curves 3 and 4 would rise uniformly, like curves 1 and 2, and would not be discontinuous. This leads us to assume that during paired stimulation, the abrupt increase in the 2nd H-response during the last 25-30 msec of the latent period is apparently determined by a depression of the spinal inhibitory system that had evoked the inhibition of the 2nd reflex response prior to that time.

Thus, the results of experiments involving the testing of the latent period by paired stimulation suggest that one of the components of the organization of voluntary movement is a depression of the spinal interneuronal inhibitory system of the agonist that is particularly intense during the last 30 msec of the latent period of the voluntary movement.

Inhibitory Interneuronal System of the Agonist That Is Activated by Antidromic Stimulation

As has been proved in experiments the results of which were discussed in Chapter 2, Section 1, the depression of the H-response accompanying an increase in the strength of nerve stimulation (H-response of zone III) is partly the result of central (intraspinal) inhibition. As an analysis of this phenomenon has shown, it is most probable that such central inhibition is determined by the activity of inhibitory Renshaw cells triggered by the antidromic activation of motoneuronal axons, or (less likely) of inhibitory interneurons activated by relatively fast-conducting afferent fibers (presumably the I-b afferents). Again, the absence of a more definite response forces us to designate the inhibitory interneurons responsible for the depression of the "supramaximal" H-response by a more general term: the spinal inhibitory system.

If there is a change in the state of this system during the organization period, it may be reflected in the character of the change in the supramaximal H-response during the latent period of voluntary movement. Experiments have shown that the amplitude of the supramaximal H-response (zone III) increases relatively little in the interval from 60 to 35 msec prior to the onset of voluntary electromyographic activity, and rises sharply during the last 30 msec immediately preceding the voluntary EMG volley (Kots, 1964; Gurfinkel' et al., 1965).

Thus, in testing the latent period of voluntary movement by means of the supramaximal H-reflex as well as the paired H-reflex, the last 60-msec interval of the latent period appears to be divided into two parts with a nonuniform "break" in the magnitude increase of the test reflex. The first part is probably associated with a steady increase in the reflex excitability of the motoneurons of the future agonist ("tuning"). In the second part of the interval (the last 25-30 msec of the latent period), a second effect appears to be superimposed on the "tuning" increase. The sharp rise in the amplitude of the supramaximal H-response during the last 25-30 msec of the latent period may be explained by the sudden depression of the inhibitory spinal system that is activated by strong nerve stimulation and evokes the inhibition of a certain portion of the GM motoneurons during rest.

This depression (the inhibition of inhibition) probably removes the effects of those inhibitory influences to which the motoneuron pool of the agonist might be subjected in connection with its impulse activity and the corresponding contraction of the innervated GM. Such influences include the inhibition of motoneurons via the Renshaw cells (antidromic inhibition), which are activated under natural conditions by impulse activity from the motor nucleus, as well as inhibition via the interneurons that are activated by the Golgi tendon receptors during muscular contraction (autogenic inhibition) (Granit, 1955).

A depression of the spinal inhibitory system of the agonist preceding the onset of movement may satisfactorily account for the abolition of the "silent period" at the onset of voluntary phasic movement. P. Hoffmann (1922) was the first to discover that a reflex H-response or Achilles tendon response elicited during voluntary GM activity is followed by a period of several dozen milliseconds in which these muscles are electrically quiescent. This phenomenon was subsequently termed the "silent period," and has been confirmed in a great many studies in humans and animals (Hoff, A. E., et al., 1934; Rusinov and Chugunov, 1945; Granit, 1955; Higgins and Lieberman, 1968).

More recent findings suggest that the "silent period" is the result of a number of factors. The refractoriness of motoneurons does not appear to play a significant role in this period, since it is observed in a muscle following direct electrical stimulation of its synergist that does not evoke the excitation of the motoneurons of the muscle itself (Granit, 1955).

Fulton and Pi-Suñer (1928) associated the silent period with a pause in muscle-spindle firing during an evoked muscular contraction. In accordance with this hypothesis, Merton (1950) found that cooling a muscle, thereby protracting contraction, lengthens the silent period.

On the basis of data obtained in the laboratory of Granit (1955), it was hypothesized that the silent period may be partly connected with the gain in impulsation from the Golgi tendon receptors during evoked muscular contraction (autogenic inhibition). By comparing the duration of the silent period electrically evoked in the human biceps brachii with and without muscle contraction, Hufschmidt (1960) also concluded that the silent period is associated with autogenic inhibition from the Golgi tendon receptors.

Because the impulse frequency from the tendon organs is a function of muscle tension, one might expect that the duration of the silent period is also a function of tension. With this in mind, Jansen and Rudjord (1964), while experimenting with cats, compared the duration of the silent period of a muscle during isometric and isotonic contractions. They found that despite the substantial differences in muscle tension from one mode of contraction to the other (from 400 g during isometric to 14 g during isotonic contraction), the duration of the silent period was the same in both cases. In the same study, it was shown that significant differences exist in the behavior of the tendon organs during isometric and isotonic contraction. On the basis of their findings, the authors concluded that the tendon organs play no significant role in determining the duration of the silent period.

It is noteworthy that the appearance of the silent period immediately after the H-response and at least the first 30 msec of the silent period are not the result of the cessation of muscle-spindle discharge or an increase of discharge from the Golgi tendon organs, because in man the minimum delay in the reflex arc is approximately 30 msec. Judging by experimental data, the first 30-50 msec of a silent period evoked by the electrical stimulation of a muscle or its nerve are evidently the result of antidromic inhibition by the Renshaw cells (Eccles, J. C., et al., 1954). As early as 1934, A. E. Hoff et al. (1934) had discovered that the duration of the silent period is unchanged in a deafferented muscle. These findings were confirmed by Struppler and Struppler (1960). They recorded the silent period in the short and long heads of the biceps and in the brachialis muscle approximately 20 msec after electrical stimulation of the long head of the biceps in a patient whose posterior roots had been sectioned from C_4 to Th_1, inclusively, for the alleviation of phantom pains following amputation of the right hand. As under normal conditions (Merton, 1951; Schoen, 1951), the duration of the silent period of the deafferented muscle was proportional to the strength of muscle stimulation and inversely proportional to background voluntary activity.

Next, it was useful to study the silent period during the performance of a voluntary phasic movement, since the presence or absence of this phenomenon and its prominence, particularly during the first 30 msec following the test H-response, may depend primarily on the state of the Renshaw cells.

First let us recall (see Section 1 above) that regardless of its magnitude and the moment of its elicitation, the test H-reflex has no effect on the initiation of voluntary motoneuron impulse activity, and may thus be observed in any interval prior to the onset of the voluntary EMG volley, including immediately before its onset. Figure 19A shows records of the H-response preceded by a strong M-response at various intervals near the onset of the EMG volley: 25 msec (1st trace), 15 msec (2nd trace), and immediately preceding (or accompanying) the onset of the EMG volley (3rd and 4th traces). Records of a supramaximal H-response preceded by an M-response with an amplitude approximately 60% of the maximal M-response are given in Fig. 19B. The first three traces show the H-responses prior to onset of the voluntary EMG volley. It can again be seen that the test H-response may be immediately followed by voluntary EMG activity with no silent period. Thus, neither the synchronous reflex discharge of a

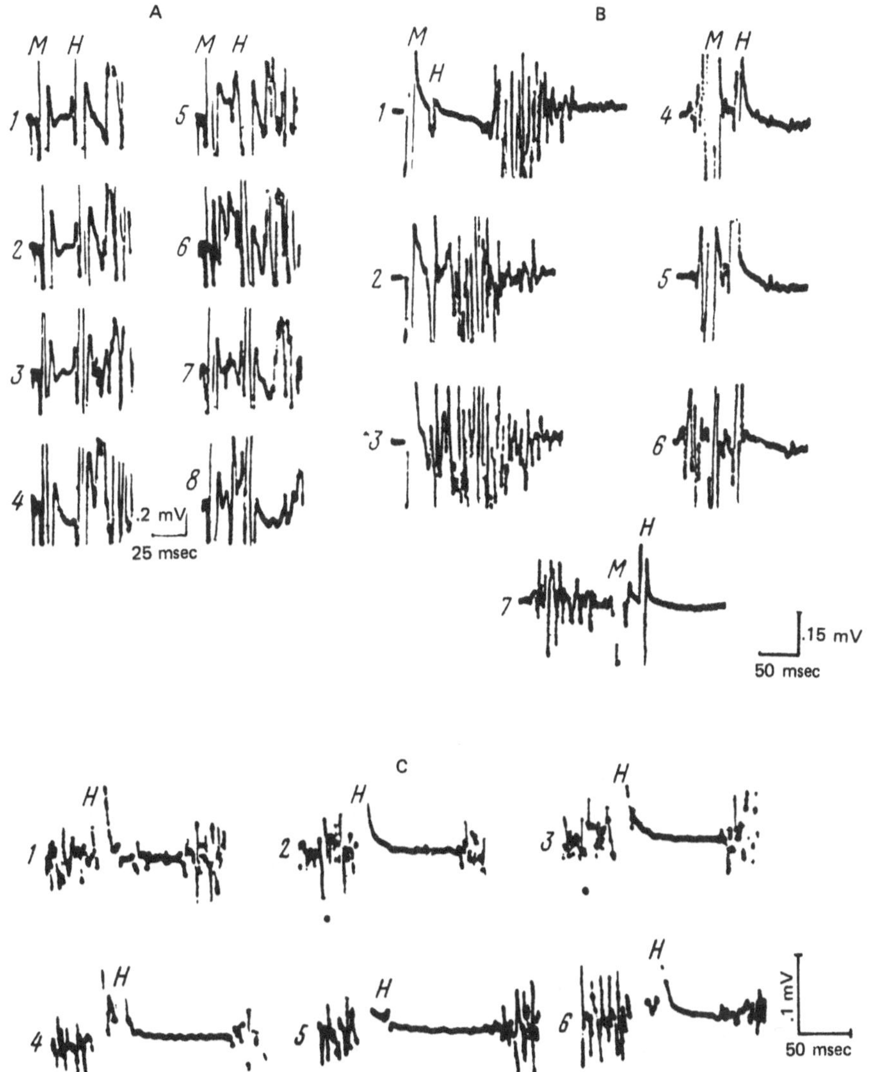

Fig. 19. "Silent" period during phasic (A,B) and static (C) voluntary contraction of GM. (A,B) Supramaximal H-responses at various intervals prior to and after onset of voluntary EMG volley of GM; (B) strength of nerve stimulation greater than A; (C) strength of nerve stimulation increases from 1st through 6th trace during slight (1-5) and strong (6) voluntary static tensioning of GM. Further explanation in text.

large number of the motoneurons of the future agonist nor a significant antidromic volley transmitted over their axons into the spinal cord evokes the inhibition of the initial voluntary impulse activity of these motoneurons.

Moreover, like other authors (Merton, 1951; Schoen, 1951), we noted that during static voluntary muscular contraction following evocation of the H-reflex, the duration of the silent period is directly proportional to the amplitude of the H-response. Records of the electrical activity of the GM during moderate static effort are shown in Fig. 19C (traces 1-5). It can be seen that as the stimulus strength increases (and, correspondingly, the amplitude of the H-response), the silent period lengthens and becomes more pronounced. A different result is observed when the H-reflex is evoked during voluntary static muscle tensing during the latent period preceding phasic movement (see Chapter 3, Section 2). Despite the steady increase in the amplitude of the test H-response as the onset of the phasic EMG volley draws near, there is no silent period, and the moment at which the H-response is evoked has no effect on the onset of voluntary EMG activity (see Fig. 12A). The latter observation indicates that during the organization of a voluntary phasic movement, the silent period of the reflexly excited motoneurons is shortened (or is entirely abolished).

The absence of an inhibitory effect after test stimulation is observed not only with respect to the onset of voluntary impulse activity of the agonist motoneuron pool. Figure 19A (traces 5-8) shows recordings of H-responses during the voluntary EMG volley: the onset of the H-response 5 msec after onset of the EMG volley is shown in trace 5; trace 6, 25 msec after EMG; trace 7, 35 msec; and trace 8, 40 msec after onset of the voluntary EMG volley. It can be seen that following an H-response elicited at the onset of voluntary impulse activity, the silent period of the motoneurons either is entirely absent or does not exceed 10 msec. The later the test stimulus is applied to the nerve after the onset of impulse activity by the agonist motoneuron pool, the more pronounced and prolonged is the silent period (see Fig. 19B, traces 4-6). If the H-response is elicited 80-100 msec or longer after the onset of voluntary activity, it completely abolishes the "tail" of the EMG volley (Fig. 19A, trace 7).

It should be noted that the greatest inhibitory effect is observed during nerve stimulation that elicits a significant M-response. An H response that is not preceded by an M-response frequently elicits no silent period at all during approximately the first 30-50 msec of the voluntary phasic EMG volley. The recordings shown in Fig. 12A (traces 11-13) indicate that the silent period is absent subsequent to the H-response elicited simultaneously with the onset of the EMG (trace 11), and approximately 15 (trace 12) and 60 msec (trace 13) after the onset of the voluntary phasic EMG volley.

Thus, a test stimulus that elicits strong M- and H-responses from an agonist at the onset of voluntary impulse activity by its motoneuron pool generally does not lead to complete inhibition or evokes a very brief period of inhibition of this activity. One of the likeliest explanations for these facts is that during the organization and onset of voluntary movement, there is an inhibition of those spinal inhibitory systems (in particular the Renshaw cells) on the activation of which the silent period depends. Such an explanation is also in agreement with our other findings, described earlier, according to which the organization of voluntary movement is accompanied by the depression of the spinal inhibitory systems of the motoneuron pool of the future agonist. The absence of the silent period appears to be one of the results of this depression.

Hufschmidt (1966) also found that during the performance of a rapid voluntary movement, electrical stimulation of the GM nerve or the GM itself does not evoke the silent period. According to Hufschmidt's findings, during such a movement, the test H-response is reproduced with a frequency up to 150/sec; i.e., the motoneurons reflexly produce a frequency that is achieved only during the artificial intracellular depolarization of motoneurons (Shapovalov, 1966). While passive flexion of the foot during static voluntary effort reduces the amplitude of the test H-response, the same procedure during voluntary movement has no effect on the amplitude and high-frequency reproducibility of the H-responses. On the basis of his findings, Hufschmidt concluded that during a rapid voluntary movement, there is an inhibition of those interneurons that are activated by impulses from the Golgi tendon organs and that have an inhibitory effect on motoneurons (interneurons of autogenic inhibition). Earlier, we presented experimental data that showed that impulses from the tendon organs play no significant role in the development of the silent period, or in any event the first 30 msec of this period.

In analyzing the data obtained, it must be remembered that the presence or absence of the silent period as well as its prominence are dependent not only on the state of the inhibitory system that determines the appearance of this phenomenon, but also on the level of motoneuronal excitability. It might well be assumed that the inhibitory effects triggered by the test stimulus are "neutralized" by powerful facilitatory effects acting on the agonist motoneurons during the preparation for and the performance of a voluntary movement. If this is the case, then the absence or near-absence of the silent period is determined not only (and possibly not so much) by the inhibition of the system that inhibits the motoneuron pool of the agonist, but also (or rather) by powerful facilitatory influences on this nucleus that mask the inhibitory effects. Two facts lead us to believe that the latter explanation is not exhaustive:

First, there is no correlation between the presence and duration of the silent period during the initial period of voluntary movement and the level of the reflex excitability of the motoneurons. As will be shown later (Chapter 4, Section 3), during the first 60 msec or so of the voluntary EMG volley, the reflex excitability of the agonist motoneurons remains more or less constant. If the absence of the silent period were determined only by a high level of motoneuron excitability, the period ought to be absent during the entire 60-msec interval. However, during this interval, the silent period changes from complete absence at the very onset of the EMG volley to a period with a duration of 20-30 msec toward the end of the interval. Evidently, the powerful depression observed during the organization period is followed by a gradual decrease in the degree of depression of the spinal inhibitory system acting on the motoneuron pool of the agonist.

Second, experiments involving the determination of the silent period during voluntary static effort have shown that even during maximum static effort, application of the test stimulus is followed by the complete electrical silence of the muscle, lasting approximately 50 msec (Fig. 19C, trace 6). This suggests at once that by itself, the powerful facilitation of spinal motoneurons cannot entirely prevent their inhibition as evoked by nerve stimulation. Moreover, the presence of the silent period during voluntary static ("tonic") activity of the motoneuron pool and its absence during the organization and at the onset of voluntary "phasic" activity of the agonist motoneuron pool allow us to assume that the state of the interneuronal inhibitory system of the agonists, particularly of the Renshaw cell system, is different for the two modes of voluntary activity.

The inhibition of the Renshaw cell system is apparently essential in ensuring the high-frequency activity of the agonist motoneuron pool that is characteristic of the performance of a rapid voluntary phasic movement or of locomotor movements (Severin et al., 1968). In both cases, the volley of the motoneuron pool consists of the short-lasting, relatively high-frequency activity of a number of concurrently (or nearly concurrently) firing agonist motoneurons (Gilson and Mills, 1941; Severin et al., 1967). The depression of the mechanism of recurrent inhibition via the Renshaw cells permits this form of motoneuronal activity. In addition, the depression of the recurrent inhibition of the agonist promotes the effective reciprocal inhibition of the antagonist, since it reduces the recurrent inhibition of the I-a interneurons of reciprocal antagonist inhibition (see Chapter 4, Section 3).

During exertion of a constant force (static effort), a different mode of pool activity is required that involves asynchronous, relatively low-frequency motoneuronal activity; it is likely that this activity is also provided for by the active participation of the mechanism of recurrent inhibition (Gel'fand et al., 1963; Gurfinkel' et al., 1964).

We believe it is quite significant that the change in the state of the spinal interneuronal inhibitory system of the agonist appears to proceed independently of the "tuning" increase in the reflex excitability of the motoneuron pool of the agonist. According to all the tests, the most significant change in the state of the interneuronal inhibitory system begins 30 msec prior to the onset of their voluntary impulse activity, and the steady increase in the reflex excitability of the motoneurons of the future agonist begins approximately 60 msec prior to the onset of their voluntary impulse activity. The sharp depression of the interneuronal inhibitory system of the future agonist does not accelerate the "tuning" increase in the reflex excitability of its motoneuron pool during the last 30 msec of the latent period preceding voluntary movement. The onset of the voluntary motoneuronal volley is soon followed by the gradual "disengagement" of the inhibitory system, while the level of motoneuronal reflex excitability remains more or less constant for a period of several dozen milliseconds.

The fact of the relative independence of the two phenomena--supraspinal "tuning" of the reflex excitability of the agonist motoneuron pool and depression of the agonist inhibitory system--allows us to associate the origin of these phenomena with the action of two different mechanisms (systems) that are activated by supraspinal command. It is conceivable that the rapid change in the state of the spinal interneuronal inhibitory system of the agonist during the 25- to 30- msec interval preceding the onset of voluntary movement is a result of the action of the same "triggering" supraspinal (perhaps pyramidal) effects that are responsible for the sharp increase in the reflex excitability of the "fast" motoneurons of the agonist (see Section 2 above).

CHAPTER 4

THE STATE OF THE SPINAL MOTONEURON POOL AND THE INTERNEURONAL APPARATUS OF RECIPROCAL INHIBITION OF THE ANTAGONIST DURING THE ORGANIZATION OF VOLUNTARY MOVEMENT

On the basis of detailed experiments performed with monkeys and cats dealing with the reflex interaction of the spinal centers of antagonist muscles, Sherrington (1906) reached the conclusion that the degree of tension of one of the muscles of a pair of antagonists has a direct effect on the tonus of the opponent--both mechanically and reflexly--over the afferent and efferent channels and the spinal cord. Sherrington then introduced the concept of "reciprocal innervation" to describe this type of interaction of a pair of antagonist muscles, thereby suggesting that reflex inhibition (relaxation) and reflex excitation (contraction) are integral parts of the same reflex reaction.

In addition to reciprocal effects for antagonist muscles during their reflex activation, analogous effects were observed during the stimulation both of supraspinal structures and of descending supraspinal pathways (Vvedenskii, 1897; Sherrington, 1906; Vvedenskii and Ukhtomskii, 1909; Ukhtomskii, 1911; Gellhorn, 1956). Applying the monosynaptic testing method (Agnew et al., 1963; Maksimova and Sverdlov, 1966; Sverdlov, S. M., and Maksimova, 1969) and recording intracellularly from individual motoneurons (Corazzo et al., 1963; Kato et al., 1964; Agnew and Preston, 1965; Vasilenko and Vucho, 1966), researchers have determined that electrical stimulation of the sensorimotor cortex or bulbar pyramids in cats elicits reciprocal effects in the motoneuron pools of flexors and extensors--the facilitation of the former and the inhibition of the latter. On the other hand, Preston and colleagues (Preston and Whitlock, 1960, 1963; Preston et al., 1967) found that in primates, a pyramidal volley evokes complex changes in the monosynaptic reflex (motoneuronal excitability): first the brief facilitation of flexor and extensor motoneuron pools, followed by inhibition (primarily of the extensors), and finally late facilitation. Thus, unlike the reciprocal pyramidal effect for flexor and extensor motoneurons observed in cats, in the primate these motoneuron populations can be classed only according to the later inhibition of extensors and facilitation of flexors following the initial general monosynaptic (direct) facilitation of all motoneuron pools.

In addition to the evocation of reciprocal effects by stimulation of the cortex of pyramidal pathways, these effects are also evoked in the spinal center of antagonist muscles during natural activity (Beritov, 1916; Magnus, 1924) and during the artificial stimulation of various subcortical structures (Brodal, 1957; Cherkes, 1963; Magoun, 1963; Amatuni, 1967; Shapovalov, 1968, 1970) and descending motor systems: the rubrospinal (Pompeiano, 1957; Sasaki et al., 1960; Shapovalov and Shapovalova, 1966; Kostyuk, 1967; Kostyuk and Pilyavskii, 1967; Shapovalov and Karamyan, 1968), the vestibulospinal (Sprague et al., 1948; Gernandt and Thulin, 1955b; Gernandt et al., 1957; Brodal, 1957; Shapovalov et al., 1966; Shapovalov and Saf'yants, 1968; Lund and Pompeiano, 1968), and the reticulospinal (Sprague and Chambers, 1954; Gernandt and Thulin, 1955a; Kostyuk and Preobrazhenskii, 1967; Limanskii and Preobrazhenskii, 1969).

Presenting arguments in support of the "spinal localization" of the phenomenon of supraspinal reciprocal inhibition of the antagonist, Sherrington (1906) pointed out that this phenomenon is observed not only during stimulation of the cortex, but also on stimulation of the white subcortical matter or the internal capsule following removal of the cortex. He concluded from this observation that inhibition received from the cerebral cortex is not determined exclusively by the interaction of cortical neurons, and possibly not at all. Vvedenskii (1913) was also inclined to believe that "the coordination of inhibition and excitation in the extremities is still essentially the province of the spinal cord" (p. 295). Summarizing experimental data dealing with stimulation of the reticular formation, Rossi and Zanchetti (1960) note that it is quite probable that the excitation or inhibition of a given motoneuron group leads to the inhibition or excitation of the antagonistic motoneurons through purely reflex or propriosynaptic mechanisms. In other words, reciprocal organization appears to be a property intrinsic to the spinal cord itself, rather than a result elicited by a reticulospinal impulse. Based on his own findings, Kryzhanovskii (1968) also concludes that descending supraspinal effects are translated into inhibitory effects only at the spinal level.

On the other hand, experiments by Lundberg and his collaborators involving transection at different levels of the spinal cord have shown that supraspinal influences can, in principle, provide for the differentiated control of each of the two components of a reciprocal reflex reaction, i.e., excitation in one group of motoneurons and reciprocal inhibition in the antagonistic group. In particular, experiments performed by Holmquist and Lundberg (1961) have demonstrated that transection at the level of the lower border of the pons in the cat leads to the increased reciprocal inhibition of extensor motoneurons in response to stimulation of flexor reflex afferents with no accompanying increase in the facilitation of flexor motoneurons. The latter is observed after a more caudal transection of the medulla oblongata. Moreover, following injury at the lower level of the pons, impulses from the flexor reflex afferents may evoke the inhibition of flexor motoneurons. Thus, when a command is issued by the higher centers, the reciprocal reflex effects in the motoneurons of an antagonistic pair are not necessarily related in the manner proposed by Sherrington in his hypothesis on the reciprocal interaction of antagonist muscles as a "physiological unit."

In fact, while utilizing intracortical microstimulation in cats, Asanuma and Sakata (1967) found within the motor cortex discrete "effective zones" that, when stimulated, produce opposite changes in the amplitude of the test monosynaptic reflex. The localization of the "effective" inhibitory zones is generally deeper in the cortex than that of the "effective" facilitatory zones. The facilitatory and inhibitory zones for different muscles usually overlap, such that the stimulation of one site in the cortex can simultaneously evoke facilitatory and inhibitory effects on the monosynaptic reflexes of different muscles, with each of the two effects having its own threshold.

It was later shown (Asanuma and Ward, 1971) that during intracortical microstimulation of the motosensory cortex in the cat at the level of the Vth and VIth layers, it is possible to evoke the contraction of individual forelimb muscles without the simultaneous (reciprocal) relaxation of the antagonists. The volume of cortex that must be stimulated to evoke the contraction of a muscle does not always coincide with the volume that must be stimulated to inhibit the contraction of the antagonist. These two zones sometimes overlap, but are often separately localized. By simultaneously stimulating different sites in the motosensory cortex, it is possible to evoke all possible combinations of the activity of an antagonistic pair: simultaneous contraction, reciprocal interaction, or the individual contraction of each muscle. Such modes of interaction have also been observed with one of the stimulating electrodes imbedded in the white matter. It could thus be assumed that the seat of the interaction in these cases is not the motor cortex, but the spinal cord (Asanuma and Ward, 1971). Since relaxation of the antagonist could be observed even without an accompanying change in the activity level of the agonist, it is not very likely that the coactivation of the gamma loop is involved in this interaction. Thus, Asanuma and Ward conclude that the motosensory cortex in the cat can control distal forelimb muscles independently of the spinal reciprocal mechanisms.

Analogous results were obtained in experiments with monkeys (Asanuma and Rosen, 1972), in which the discrete spatial distribution of excitatory and inhibitory zones for each muscle in the motor cortex was also observed. V. B. Brooks and Stoney (1971) suggest that the use or nonuse of segmental reflex mechanisms (e.g., reciprocal inhibition) by the corticofugal motor output is a function in which groups of corticofugal neurons are active.

It is known that an extremely wide variety of interaction modes are displayed by antagonist muscles during the performance of different natural movements--from purely reciprocal (according to Sherrington's concept) to simultaneous contraction, depending on the speed, external load, and other parameters of movement (Wilson, S. A. K., 1929; Person, 1965). It is the opinion of Person (1965) that these facts tend to refute "the notion that reciprocal inhibition plays a predominant role in natural motor activity" (p. 42).

At the same time, a number of studies have provided direct evidence of reciprocal inhibitory action on the motoneuron pool of the antagonist during voluntary agonist activity. For example, P. Hoffmann pointed out as early as 1922 that the voluntary tensioning of a foot flexor group inhibits an electrically evoked (H-) reflex or the Achilles reflex of the antagonistic foot extensors. This observation, first made by P. Hoffmann, has since been repeatedly confirmed (Wachholder and Altenburger, 1927; Paillard, 1959; Gurfinkel' et al., 1965; Mayer and Mawdsley, 1965; Kots, 1969a,b; Baranov-Krylov, 1969; Gottlieb and Agarwall, 1972). According to the findings of Baranov-Krylov, the depression of excitability of antagonist motoneurons during voluntary agonist activity, i.e., the degree of reciprocal inhibition, is proportional to the level of agonist activity.

The reflex activation of motoneurons in man leads to the reciprocal inhibition of the background voluntary activity of the antagonist motoneurons (Peimer and Perli, 1950; Kugelberg and Hagbarth, 1958). The tonic "after"-activity of the foot extensors evoked by subthreshold tetanization of the TN is inhibited during voluntary tensioning of the antagonistic flexors. After the cessation of voluntary tensioning of the antagonists, the evoked activity of the extensors is restored (Lavy and Valbo, 1967).

A number of investigators agree in their assumption that the great variety of the interrelationships between antagonist muscles observed in the course of natural motor activity is due to the participation of the higher levels of the central nervous system. However, the concrete mechanisms of this participation during the performance of voluntary movements have not yet been studied. The goal of our investigation was to study the spinal mechanisms of the voluntary control of antagonist muscles in man. The first step in solving this general problem was to investigate the dynamics of the reflex excitability of the motoneuron pool of the antagonist during the latent period, at the onset of a phasic voluntary movement, and during the performance of a sequence of two movements in opposite directions, in which case the activity of the antagonist muscles is alternated. The results of these experiments are discussed in Sections 1 and 2. Section 3 presents data taken from the literature and obtained by the author on the supraspinal regulation of the interneuronal spinal apparatus involved in the reciprocal inhibition of antagonists.

1. THE MOTONEURON POOL OF THE ANTAGONIST

To investigate the supraspinal effects on the motoneuron pool of the future antagonist of a voluntary movement, the H-reflex test method was employed during the latent period and at the onset of dorsal flexion of the foot, a movement in which the GM assumes the role of the primary antagonists.

Solitary Movement (Flexion)

As was mentioned earlier (Chapter 3, Section 1), at the onset of the latent period of any voluntary movement, the reflex excitability of the GM motoneurons is already higher than at rest. In Table 6, changes in the amplitude of the test H-response of the Sol during the first half of the latent period of two voluntary movements--extension and flexion of the foot--are compared. In these movements, the Sol plays the role of the agonist and the antagonist, respectively. The appearance of the beam on the oscilloscope screen served as the movement signal. The amplitude of the control H-response was 20-30% of the maximal. For a more precise determination of the onset of voluntary EMG activity by the agonist, the recordings on one of the channels were made at high amplification (see track III in Fig. 20A and track I in Fig. 20B). The other channel was used to record the test H-response of the Sol (track II in Figs. 20A and B). During flexion of the foot, the interval from the H-response of the ATM to the onset of the voluntary EMG volley of the ATM was determined by the EMG records of the ATM made at high amplification (see track I in Fig. 20B). In each experiment, the averaging of the amplitude of the H-response in a given interval prior to the EMG volley was based on no fewer than 5 responses. The average number of H-responses falling within a given interval in each experiment is designated in Table 6 and all subsequent tables as n. The number of experiments in which calculations were made of the amplitude of the H-response in a given interval (cases in which the number of H-responses was ≥ 5) is designated as N.

As shown by the data presented in Table 6, during the early phase of the latent period of voluntary movement, the reflex excitability of the motoneuron pool of the future antagonist (of foot flexion) is higher than at rest, but this "background" increase (see Chapter 3, Section 2) is less than the "pretuning"

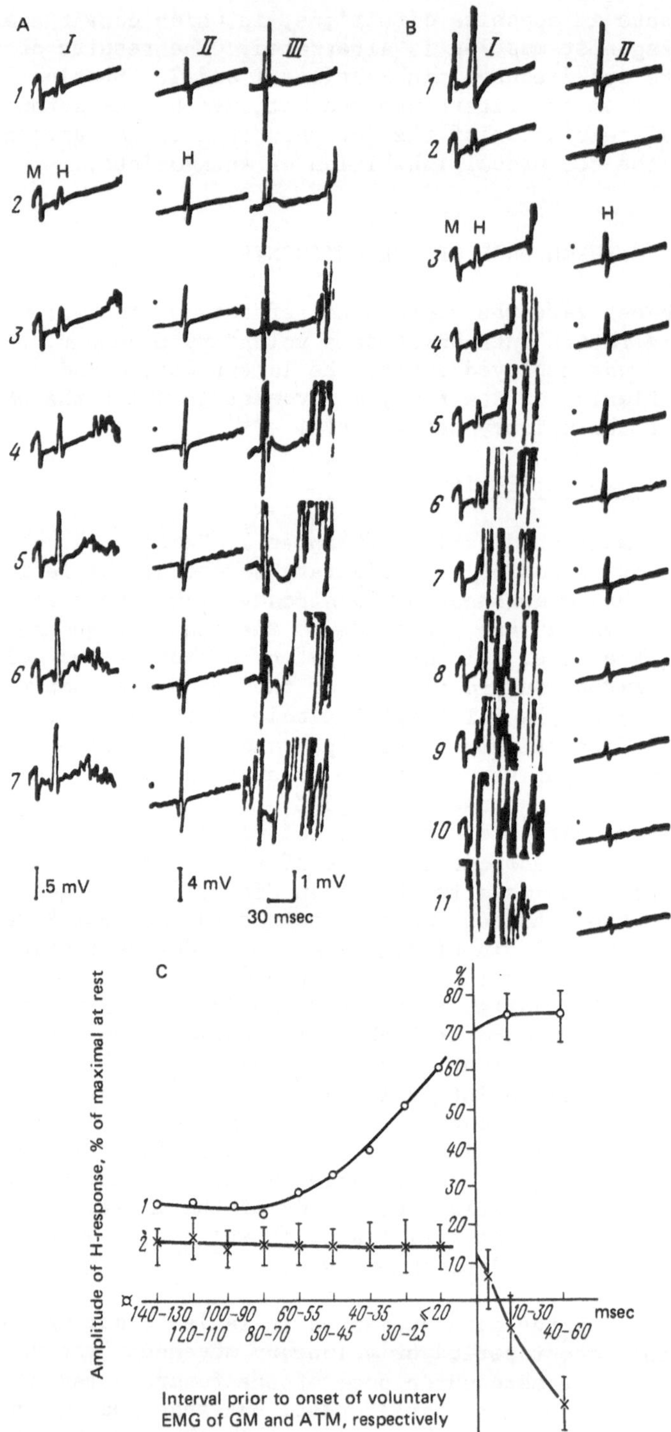
A
I
II
III
M H
H
.5 mV
4 mV
1 mV
30 msec
B
I
II
M H
H
C
Amplitude of H-response, % of maximal at rest
%
80
70
60
50
40
30
20
10
140-130
120-110
100-90
80-70
60-55
50-45
40-35
30-25
≤20
10-30
40-60
msec
Interval prior to onset of voluntary
EMG of GM and ATM, respectively

Table 6. Average Amplitudes of H-Response of Sol During Early Phase of Latent Periods of Foot Extension and Flexion (in % of Control Amplitude at Rest)

Index	Interval from H-response to onset of voluntary EMG of agonist (msec)				
	140-130	120-110	100-90	80-70	60-50
	Extension of foot				
M	125.0	122.3	129.1	122.7	140.0
±m	±12.5	±6.7	±6.7	±6.9	±19.3
N	7	12	12	10	5
n	9.0	12.0	22.0	15.2	7.3
	Flexion of foot				
M	115.1	117.3	111.4	114.5	114.3
±m	±5.1	±6.5	±4.5	±7.4	±11.2
N	7	11	11	12	7
n	7.5	10.2	14.7	17.6	4.6

Fig. 20. H-response of Sol during latent period and at onset of voluntary extension (A) and flexion (B) of foot. (A) (I) EMG of ATM (beam sweep beginning from M-response); (II) EMG of Sol (sweep beginning from shock artifact); (III) EMG of MG (sweep beginning from H-response); (1-6) H-response as onset of voluntary EMG volley of GM draws near; (7) H-response at onset of voluntary EMG volley of GM. (B) (I) EMG of ATM; (II) EMG of Sol; (1) maximal H-response of Sol; (2) control H-response of Sol at rest; (3-7) H-response as onset of EMG volley of ATM draws near; (8-11) at onset of voluntary EMG volley of ATM. (C) Change in amplitude of test H-response of Sol prior to extension (1) and flexion (2) of foot.

increase in the reflex excitability of the motoneuron pool of the future agonist (of foot extension). Thus, even early changes in motoneuronal excitability are "specific" in the sense that though the state of attention, expectancy, and readiness in the motoneuron pool of both the agonist and the antagonist evokes an increase in reflex excitability, the degree of this increase is significantly higher in the motoneuron pool of the agonist of the future movement than in its antagonist. On the average, in the interval 140-70 msec before the onset of voluntary movement, the amplitude of the test H-response of the Sol amounts to 123.6±3.8% of the resting magnitude (110.0-156.0%) prior to foot extension and 113.3±3.4% (87.0-129.1%) prior to foot flexion.

As described earlier (Chapter 3, Section 2), the level of the "pretuning" increase in the reflex excitability of the agonist motoneuron pool remains constant only during the first ("pretuning") phase of the latent period. During the last 60 msec of the latent period, there is a gradual increase in the reflex excitability of the motoneuron pool of the future agonist (the second, "tuning" phase of the latent period). Unlike that of the agonist, the heightened reflex excitability of the motoneuron pool of the antagonist of the future movement remains unchanged throughout the entire latent period. As shown by the data in Table 6, this assertion holds true for the entire latent period up to the last 50 msec prior to the onset of voluntary impulse activity of the motoneurons of the antagonist. That the assertion is just as true for the last 50 msec of the latent period is shown by the data from 46 systematic experiments presented in Table 7.

As the data in Table 7 indicate, during the entire latent period (except for the 10- to 0-msec interval), the amplitude of the test H-response of the future antagonist remains constant. In addition, the amplitude in the 20-msec interval does not differ significantly from the amplitude of responses in earlier intervals. The first significant decrease in the amplitude of the test H-response of the antagonist occurs in the 10- to 0-msec interval of the latent period. This is how we distinguish those cases in which the voluntary EMG volley of the ATM immediately follows the H-response of this muscle (see trace 6 in Fig. 20B). Since the duration of the H-response of the ATM is approximately 10 msec, in the latter case it is impossible to ascertain whether the test H-response occurred during the last 10 msec of the latent period or whether it accompanied the onset of the EMG volley of the antagonist involved in the voluntary movement. With this 10-msec correction, it may be concluded that the reflex excitability of the motoneuron pool of the future antagonist remains unchanged throughout the entire latent period. The absence of any significant signs of inhibitory effects on the motoneuron pool of the antagonist during the latent period is particularly noteworthy.

Table 7. Average Amplitudes[a] of H-Response of Antagonist (Sol) During Latent Period and at Onset of Flexion of Foot

Index	Interval from onset of Sol H-response to onset of EMG volley of ATM (msec)						Interval from onset of EMG volley of ATM to H-response of Sol (msec)	
	≥60	50	40	30	20	10–0	10–30	40–60
M	100	97.6	98.6	96.9	93.0	76.1	61.1	43.2
$\pm m$	--	$\pm$9.4	$\pm$7.8	$\pm$6.7	$\pm$7.9	$\pm$15.5	$\pm$14.0	$\pm$17.9
N	46	18	22	14	20	30	25	33
n_{avg}	20.8	13.2	17.2	15.5	13.2	18.8	20.2	24.6
$n_{min-max}$	10–63	10–25	10–35	10–28	10–19	10–35	10–38	10–55

[a]Here and in future, the amplitude of H-responses in intervals of 60 msec and more is assumed to be 100%.

Accompanying (accuracy $\pm$10 msec) the onset of the voluntary impulse activity of the agonist motoneuron pool is a sharp decrease in the reflex excitability of the antagonist motoneuron pool. This decrease can be considered a manifestation of the inhibition of the motoneuron pool of the antagonist. During the first 60 msec of the voluntary EMG volley of the agonist, the amplitude of the test H-response of the antagonist declines quite uniformly (see the graph in Fig. 20C).

Thus, the following conclusion may be drawn by comparing the changes in the amplitude of the test H-response of the Sol prior to two phasic voluntary movements in which this muscle acts first as the primary agonist and then as the primary antagonist and vice versa. The preparation to perform a voluntary movement in response to a signal is accompanied by simultaneous but differing increases in the reflex excitability of the motoneurons of the antagonist muscles--a stronger ("pretuning") increase in the excitability of the motoneuron pool of the agonist, and a weaker ("background") increase in the motoneuron pool of the antagonist. The level of increased motoneuronal excitability in the antagonist remains unchanged throughout the entire latent period (though the last 10 msec are indefinite), while the corresponding level in the agonist remains constant only up to the last 55-60 msec of the latent period, during which time the reflex excitability of its motoneuron pool increases uniformly and significantly. At the onset of voluntary impulse activity by the motoneuron pool of the agonist, its reflex excitability increases even more rapidly, while the reflex excitability of the motoneuron pool of the antagonist is sharply depressed (reciprocal inhibition). The effect of the reciprocal inhibition of the antagonist motoneurons increases steadily during the first 60 msec of the voluntary impulse activity of the agonist motoneurons.

Successive Paired Movements (Flexion-Extension, Extension-Flexion)

To investigate the organization of the successive activity of the motoneuron pools of antagonist muscles with the aid of the H-reflex, we determined the reflex excitability of the motoneuron pool of the GM during the performance of paired movements: flexion-extension or extension-flexion of the foot. In such movements, this muscle group alternately assumes the role of antagonist-agonist in the first case, and agonist-antagonist in the second.

The organization of the second movement of a pair of opposing movements is determined by two events: (1) the cessation of activity by the motoneuron pool of the agonist of the first movement, and (2) the initiation of activity by the motoneuron pool of the agonist of the second movement (the antagonist of the first movement). To study the two events separately, the changes in the test H-response of the Sol occurring during and after completion of the EMG volley of the first movement were determined in each experiment up to the onset of the EMG volley of the second movement. In this case, the curve describing the change in the amplitude of the H-response reflects both events. In addition, in the same experiment, the changes in the amplitude of the H-response during and after completion of the EMG volley of the first movement were determined when only the first movement was performed; in this case, the curve showing the change in the amplitude of the H-response reflects only the first event. On the basis of the difference between the two curves obtained during the same experiment, we drew conclusions regarding the spinal changes associated with the organization of the second movement of the pair.

In its resting position, the subject's foot was supported by a pedal set at an angle of 30° to the floor. Extension of the foot was very slightly opposed by the elastic action of the pedal, and during flexion the subject lifted his toes from the surface of the pedal. After solitary extension, the foot was returned passively to its initial position by the elastic force of the pedal, and after solitary flexion, by the weight of the foot itself. A paired movement consisted of the rapid alternation of the two movements in succession--flexion followed immediately by extension (flexion-extension), or extension followed by flexion (extension-flexion).

To evoke the H-reflex at different times during the performance of a movement, we utilized a special semiconductor circuit that made it possible to start the stimulator by the electrical activity of the agonist of the first movement. Briefly, the electrical activity taken from this muscle was amplified and rectified in the circuit, thereby forming at the circuit output a signal that closed an electromagnetic relay. In special experiments, we found that for the amplification factor chosen, the output relay of the semiconductor circuit was closed an average of 60 msec after the onset of the EMG volley of the first movement. The closing of the relay energized a delay unit that in turn switched on the stimulator for evoking the H-reflex. While varying the delay, we tested the reflex excitability of the spinal motoneurons of the extensors at different times: prior to and after completion of the EMG volley of the first movement, and prior to and after the onset of the EMG volley of the second movement. The strength of the nerve stimulus was kept constant throughout the experiment while both solitary and paired movements were performed.

The electrical activity of the primary antagonist muscles during the performance of flexion and extension of the foot has the following characteristics: During flexion, the EMG volley of the ATM is accompanied by some GM activity, i.e., by "concomitant activity" of the antagonists (Figs. 21A and B). The duration of the EMG volley of the ATM is, as a rule, greater during solitary

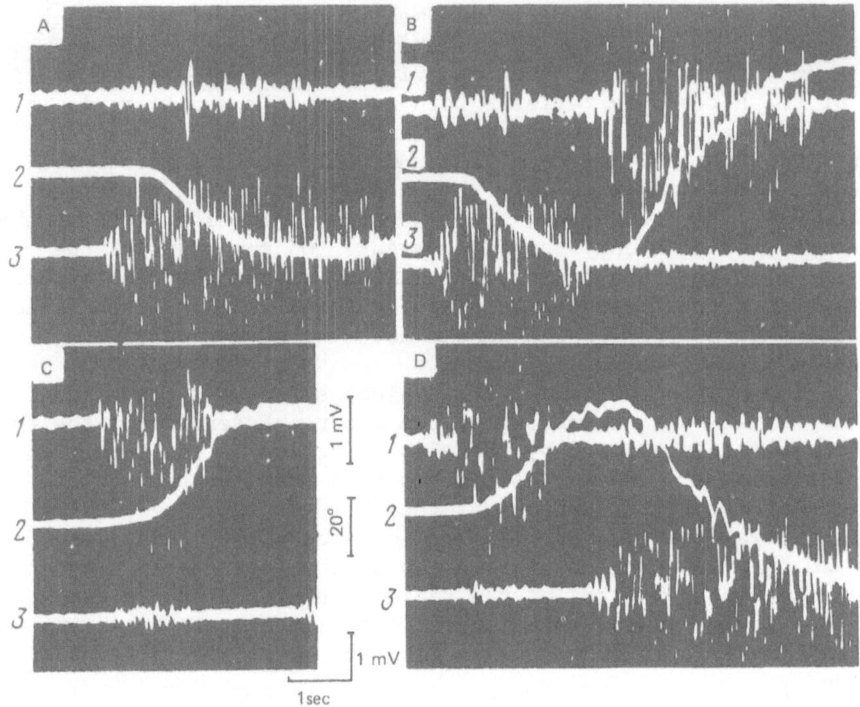

Fig. 21. EMG of antagonist GM (1) and ATM (3) and goniogram of ankle joint (2) during flexion (A), flexion-extension (B), extension (C), and extension-flexion (D) of foot.

flexion than during flexion as part of a paired movement (cf. A and B in Fig. 21). During extension of the foot, the substantial volleying activity of the GM is accompanied by some activity of the ATM, i.e., "concomitant activity" of the antagonist (Figs. 21C and D). The duration of the EMG volley of the GM is usually greater during solitary extension of the foot than during extension as part of a paired movement (cf. C and D), but is shorter than the duration of the EMG volley of the ATM during solitary flexion (cf. A and B with C and D).

During flexion-extension, the interval from the end of the EMG volley of the ATM to the onset of the EMG volley of the GM varied over a narrow range in 16 of 19 subjects: 0-30 msec (most often from 0-20 msec). Strong electrical activity of the GM during the second movement is accompanied by weak "concomitant activity" of the antagonist (ATM). During extension-flexion, the interval from the conclusion of the EMG volley of the GM to the onset of the EMG volley of the ATM varies appreciably in different subjects (from 0 to 70 msec). The variability range of the duration of the GM-ATM interval when the extension-flexion sequence is repeated by the same subject is also greater than that of the ATM-GM interval during flexion-extension. In most cases, the GM-ATM intervals are from 20 to 40 msec. Strong electrical activity of the ATM during the second movement is accompanied by weak activity of the GM.

During voluntary flexion of the foot, the test H-response of the Sol (antagonist) is sharply reduced below the resting amplitude. Therefore, in testing the reflex excitability of the Sol motoneurons during flexion and flexion-extension of the foot, the strength of the stimulus applied to the TN must be adequate to elicit a maximal H-response during rest. During the voluntary EMG volley of the ATM (agonist of the flexion movement), the amplitude of such a control H-response is reduced by approximately 50% and does not change during the last 70-80 msec of the EMG volley of the ATM. After the end of the voluntary EMG volley of the ATM during the performance of a solitary flexion movement, the amplitude of the control H-response of the Sol begins to decline noticeably (Fig. 22A, curve I). During flexion-extension, a change in the same control H-response of the Sol is observed during the last 50-60 msec preceding the onset of the voluntary EMG volley of the GM (agonist of the second movement), i.e., still during the EMG volley of the ATM (agonist of the first movement) (Fig. 22A, curve II). More frequently, the change begins with a transient decline in the amplitude of the H-response of the Sol, followed by a period marked by a gradual increase in the amplitude of the H-response as the onset of the EMG volley of the GM draws near. At the onset of the EMG volley of the GM, the amplitude of the H-response of these muscles exceeds the maximal control H-response at rest (as in conditions of solitary extension).

During the performance of flexion-extension movements of the foot, the changes in the test H-response of the Sol exhibit a time connection with the onset of the EMG volley of the GM (agonist of the second movement), and are largely independent of the time of the cessation of the EMG volley of the ATM (agonist of the first movement). This can be seen most clearly in experiments in which the interval between the cessation of the EMG volley of the ATM and the onset of the EMG volley of the GM varies over a relatively wide range. For example, in one of the experiments, this interval was 40-45 msec in 8 cases, 30-35 msec in 14 cases, 20-25 msec in 46 cases, 10-15 msec in 44 cases, and 0-5 msec in 25 cases. Figure 22B gives a graph relating the amplitude of the H-response of the Sol to its position relative to the end of the EMG volley of the ATM for different intervals prior to the onset of the EMG volley of the Sol. It can be seen that the amplitudes of the H-responses of the Sol are different for the same position relative to the end of the EMG volley of the ATM, being dependent on the interval preceding the onset of their EMG volley (EMG of GM).

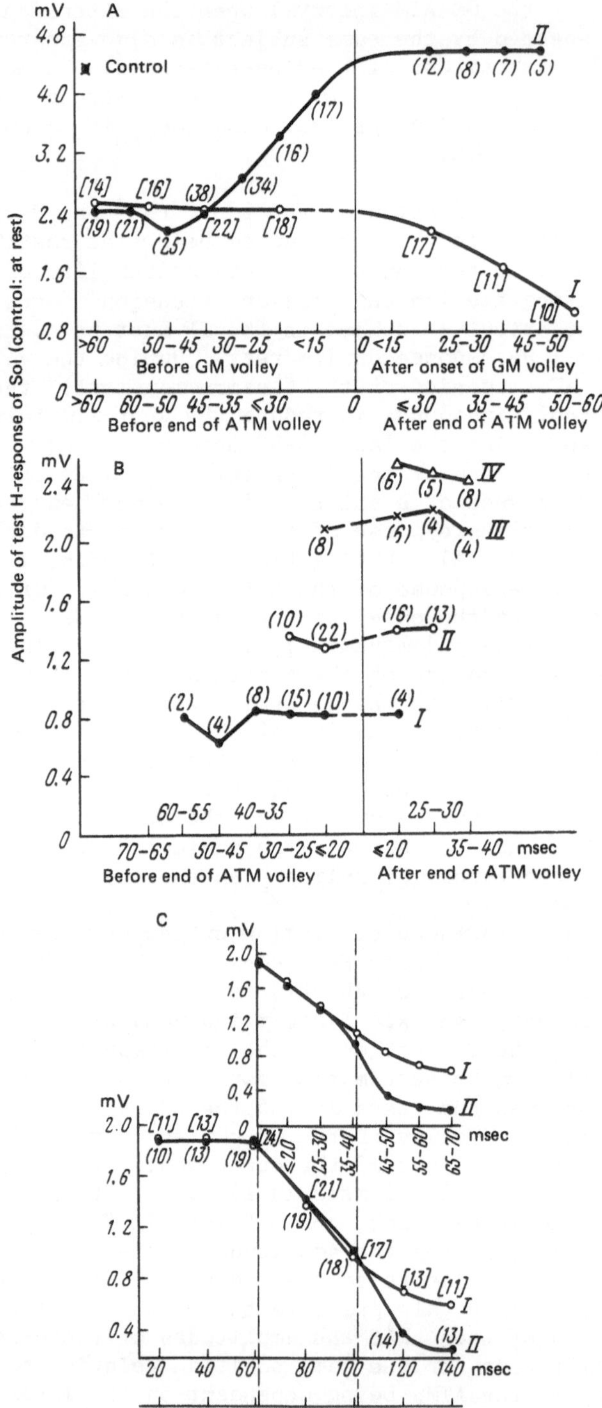
A
B
C
mV
Control
Amplitude of test H-response of Sol (control: at rest)
Before GM volley
After onset of GM volley
Before end of ATM volley
After end of ATM volley
msec

In testing the reflex excitability of the GM motoneurons during the performance of extension and extension-flexion movements, a stimulus strength was used that was adequate to elicit an H-response from the Sol at rest with an amplitude on the order of 60-80% of the maximal. During the voluntary EMG volley of the GM (agonist of the movement), such a stimulus elicits from the Sol an H-response that exceeds the maximal H-response during rest. During extension of the foot, an H-response evoked from the Sol near the end of its voluntary EMG volley is followed by a silent period in the electrical activity of this muscle. It is therefore impossible to determine the true position of the test H-response of the Sol relative to the end of the EMG volley. The absence of a gap between the voluntary EMG volley of the Sol and its H-response is evidence that the reflex response of the Sol motoneurons coincides with the period of their voluntary activity, and is not evoked after cessation of this activity. To compare the amplitudes of the H-responses at equal times after the end of the EMG volley of the GM during the performance of single and paired movements, we allowed for differences in the average duration of the voluntary EMG volley of the muscles during these two movements. For example, in the experiment the results of which are shown graphically in Fig. 22C, the average duration of the EMG volley of the GM was about 20 msec greater during the performance of a single foot extension than during extension performed as part of a paired movement: 110.0±6.3 msec and 132.0±7.8 msec, respectively. Thus, in this experiment, we compared the

---

Fig. 22. Changes in amplitude of test H-response of Sol during solitary and paired movements of the foot. (A) Flexion (I) and flexion-extension (II) of foot. Upper abscissa (for curve II): time relative to onset of voluntary EMG volley of GM; lower abscissa (for curve I): time relative to conclusion of EMG volley of ATM. (■) Amplitude of test H-response of Sol at rest. (B) Flexion-extension of foot. Abscissa: time relative to conclusion of EMG volley of ATM. Curves I-III: amplitudes of H-response prior to EMG volley of GM by 45-60 (I), 25-40 (II), and 20 (III) msec and less; curve IV: amplitude of H-response during EMG volley of GM. Numbers in parentheses indicate numbers of H-responses on which averages are based. (C) Extension (I) and extension-flexion (II) of foot. Top abscissa: interval after conclusion of EMG volley of GM; bottom abscissas--upper (for curve I) and lower (for curve II): delays of testing stimulus. Left vertical dashed line: average time of cessation of voluntary EMG volley of GM; right line (passing through both diagrams): average starting time of voluntary EMG volley of ATM. Numbers in parentheses and brackets indicate numbers of H-responses in the intervals.

amplitudes of H-responses elicited 20 msec later during the performance of a solitary movement than during the performance of a paired movement (see shift of abscissas in Fig. 22C).

During the performance of only the first movement (extension of the foot), the strong control H-response of the Sol does not change noticeably during the voluntary EMG volley of this muscle and gradually decays after conclusion of this volley (Fig. 22C, curve I). The same changes in the H-response of the extensor are observed after foot extension during the performance of the paired movement extension-flexion up to the onset of the voluntary EMG volley of the ATM (agonist of the second movement) (Fig. 22C, curve II). Accompanying (accuracy of measurement no better than 10 msec; see methodological comments above) the onset of the EMG volley of the ATM (flexor) is a sharp decline in the test H-response of the extensor. During the first 40-60 msec of the EMG volley of the ATM (observation period), the amplitude of the H-response of the Sol exhibits a more or less uniform decrease.

Thus, the results obtained in this series of investigations have shown that the changes in the reflex excitability of the spinal motoneuron pools of antagonist muscles that correspond to the organization of their successive voluntary impulse activity are similar in principle to the changes accompanying the organization of their "initial" activity following a state of rest. As before the onset of voluntary impulse activity by the Sol (agonist) during the performance of extension only, during flexion-extension of the foot, the onset of the voluntary impulse activity of the motoneuron pool of this muscle, which immediately follows the voluntary impulse activity of its flexor antagonists, is preceded by a period of a "tuning" change in the reflex excitability of the future agonist of the [second] voluntary movement of the pair. Just as before flexion only, during the performance of a paired movement (extension-flexion) of the foot, the reflex excitability of the motoneuron pool of the future antagonist (Sol) remains constant throughout the entire latent period preceding the onset of the second movement. More precisely, for extension-flexion, the test H-response of the Sol exhibits the same changes during the interval from the end of the voluntary EMG volley of the GM to the onset of the voluntary EMG volley of the ATM as the same test H-response following extension only.

In light of the latter fact, it may also be concluded that the cessation of the voluntary impulse activity of the motoneuron pool of the agonist involved in the first of a pair of rapid successive movements is not associated with the inhibitory actions on it as the future antagonist of the second movement. It may thus be assumed that this cessation is the result of the suspension of facilitatory effects on the motoneuron pool. One of the mechanisms of this suspension may be the reciprocal inhibition of the cortical center of this muscle associated with the onset of the impulse activity of the cortical nucleus of the future agonist. According to this hypothesis on the reciprocal interaction of the cortical centers of antagonist muscles, the suspension of facilitatory effects on the spinal pool of the agonist of the first movement must be directly associated with the onset of facilitatory effects on the spinal pool of the future agonist of the (second) movement--which is not consistent with our findings (see Fig. 22B). However, one must allow for the substantial time dispersion of supraspinal facilitatory effects, as well as for the probable participation of other sources of facilitatory effects on the active motoneuron pool (e.g., via the gamma loop).

When these allowances are made, it becomes clear why there is no direct connection between the time of the complete cessation of voluntary impulse activity by the motoneuron pool of the agonist of the first movement and the onset of the impulse activity of the agonist of the second movement. Significant variations in the intervals between the complete cessation of the EMG activity of the future antagonist and the onset of the EMG activity of the agonist have also been noted by other authors, particularly Koz'myan (1967, 1968) during a careful study of the Wachholder-Hufschmidt phenomenon (Hufschmidt and Hufschmidt, 1954), whereby the EMG activity of the antagonist is extinguished prior to movement.

It should be stressed that during the performance of paired voluntary movements with rapid alternation of the motoneuronal activity of the antagonist muscles, the organization of the second of a pair of movements (notably the "tuning" of the reflex excitability of the motoneuron pool of the future agonist) occurs while the first movement is still in progress. Judging by the duration of the spinal manifestations of this organization, it may be concluded that the latter is made possible by supraspinal influences similar to those accompanying the organization of a solitary movement.

Variations of the Solitary Movement (Flexion)

In this section, we shall describe the results of experiments designed to investigate the character of the interaction between the motoneuron pools of the antagonist muscles of the ankle joint during the performance of various movements in this joint (Zhukov, 1971). We chose three variations of a rapid flexion of the joint that are fundamentally different in terms of the electrical activity of the primary antagonist muscles of the ankle joint--the GM and the ATM: (1) unresisted flexion of the foot (the "standard movement"); (2) flexion of the foot against the resistance of an elastic band; and (3) flexion of the foot against a 2-kg weight. In each of these movements, the GM assumes the role of the primary antagonist; the ATM, that of the primary agonist. During the performance of these movements, electrical activity was recorded from the following muscles: the ATM, Sol, m. peroneus longus, m. extensor digitorum longus, m. biceps femoris, m. rectus femoris, and m. vastus lat.

During flexion of the foot, movement begins only about 40 msec after the onset of the EMG volley of the ATM, the primary agonist (see Fig. 21). It may thus be concluded, first, that the reflex effects from proprioceptors associated with muscular contractions and changes in muscle length, as well as with joint movements, can be felt no earlier than 70 msec after the onset of voluntary motoneuronal impulse activity. It is therefore likely that the character of the voluntary impulse activity and reflex excitability of the motoneuron pools during the first 60 msec of the EMG volley of the muscles is the result of intracentral influences that are supraspinal in origin. Second, the electromechanical delay makes it possible to test the reflex excitability of the Sol motoneuron pool during the first 40-50 msec of the voluntary EMG volley of the antagonist in the absence of any movement in the joint; this naturally precludes the possibility of artifacts associated with the displacement of the stimulating and recording electrodes. These considerations led us to evaluate quantitatively the electrical activity of the antagonist muscles only for the first 60 msec of their voluntary activity, and to test the reflex excitability of the motoneuron pool of the Sol (antagonist) only during the first 60 msec of the voluntary EMG volley of the agonist.

During the performance of all three variations, the voluntary EMG volley of the ATM, m. extensor digit. long., Sol, and m. peroneus longus begins simultaneously in all these muscles and lasts approximately 130-160 msec. There is no detectable activity of the antagonist muscles of the knee joint--the biceps femoris and quadriceps femoris muscles. Taking the electrical activity of the antagonists and their ratio as unity during the performance of the "standard movement," we find that the electrical activity of the ATM (agonist) during the first 60 msec of the EMG volley is roughly identical during the performance of all three movements: 1.0 during the "standard movement," 1.07 during flexion against the elastic resistance, and 1.10 during flexion against the weight. The electrical activity of the Sol is greater during flexion against resistance (1.20) than during the "standard movement," and is greatest during flexion against weight (1.32). Correspondingly, the smallest electrical activity ratio of these antagonists is observed during the performance of the latter movement (0.83); the ratio is higher during the second movement (0.90), and maximal during the "standard movement" (1.0). It is also noteworthy that the electrical activity of the other agonist--extensor hallucis longus--is higher during flexion against elastic resistance or weight than during the "standard movement."

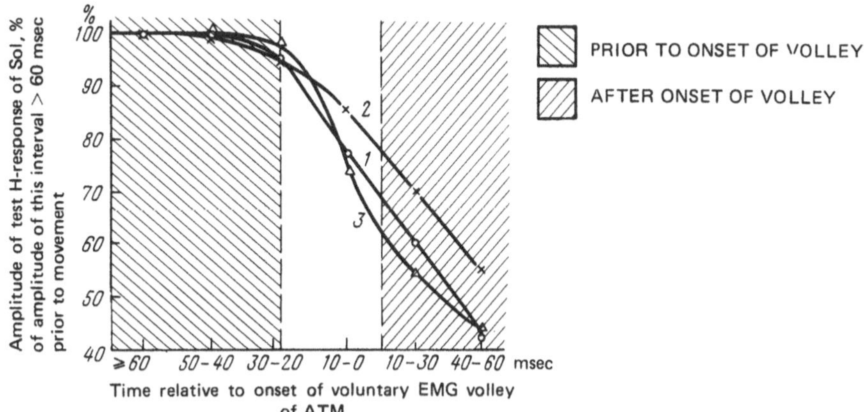

Fig. 23. Change in amplitude of test H-response of Sol during latent period and at onset of different variations of voluntary flexion of foot: (1) unresisted flexion ("standard movement"); (2) flexion against elastic resistance; (3) flexion against 2-kg weight.

Thus, during the performance of different variations of the rapid voluntary flexion of the foot, the impulse activity of the motoneuron pools of both the primary synergists--the foot flexors--and the antagonists ("concomitant activity") is recorded. The ratio of the impulse activity levels of the motoneuron pools of the antagonist muscles--the ATM and Sol--are different during the performance of the different variations of the flexion movement, at least during the first 60 msec of the motoneuronal activity of the antagonists.

The results obtained by testing the reflex excitability of the motoneuron pool of the Sol (antagonist) during the latent period and the first 60 msec of the voluntary EMG volley during the performance of different variations of the flexion movement are shown graphically in Fig. 23. As the graph shows, during the entire latent period preceding the voluntary movement, the amplitude of the test H-response of the Sol (antagonist) remains constant (the decline in the amplitude of the H-response of the Sol in the interval 30-20 msec prior to the "standard movement" and flexion against the elastic resistance is statistically insignificant). Accompanying (accuracy ± 10 msec; see above) the onset of the voluntary impulse activity of the agonist motoneurons is a sharp decline in the amplitude of the test H-response of the Sol (antagonist). This decline increases quite uniformly during the first 60 msec of the voluntary EMG volley. Since the decrease in reflex excitability is observed in the motoneuron pool of the antagonist of the movement, and it is confined to the onset of voluntary impulse activity by the motoneuron pool of the agonist, we regard this decline as a manifestation of reciprocal inhibitory effects on the motoneuron pool of the antagonist.

As was mentioned at the beginning of this section, experimental data obtained in acute experiments with animals indicate that during the stimulation of various supraspinal motor centers or their descending pathways, the facilitation of one functional group of motoneuron pools (flexors or extensors) is accompanied by reciprocal inhibitory effects in the other antagonistic functional group (extensors or flexors). In such cases, the delay from facilitation to reciprocal inhibition is only a few milliseconds. In the light of these experimental results, it might be expected that in our case the significant facilitation ("tuning") of the motoneuron pool of the future agonist beginning 60 msec prior to voluntary movement would be accompanied by the reciprocal inhibition of the motoneuron pool of the future antagonist. However, monosynaptic excitability testing of the motoneuron pool of the future antagonist did not reveal the anticipated reciprocal inhibition during the whole of the latent period (only the last 10 msec of this period are indefinite). Thus, the effects of natural supraspinal influences contrast strongly with effects evoked by synchronous electrical stimulation.

However, if we remain within the framework of the reciprocal pattern revealed in acute experiments with animals, our findings may be explained by a single, very probable assumption. First, it is significant that besides the impulse activity of the motoneuron pool of the agonist, a (somewhat weak) "concomitant" activity is recorded from the motoneuron pool of the antagonist during the performance of a voluntary movement. This suggests that the onset of this "concomitant" activity of the antagonist motoneuron pool is preceded by a supraspinal "tuning" (facilitation) of this pool that is equal in duration to that of the agonist motoneuron pool. This supraspinal facilitatory effect on the antagonist motoneuron pool during the latent period "neutralizes" its reciprocal inhibitory effect: the increase in the facilitatory "tuning" effect during the last 60 msec of the latent period is paralleled by an increase in the reciprocal inhibitory effect, such that the reflex excitability of the motoneuron pool of the antagonist remains, on the whole, relatively unchanged throughout this period.

However, a more probable assumption is that in and of itself, the supraspinal command during the organization of voluntary movement determines only the facilitatory influence addressed exclusively (or nearly so) to the motoneuron pools of the agonists of the future movement, but does not provide for the simultaneous (reciprocal) inhibitory influences on the motoneuron pools of the future antagonists of this movement. According to our findings, the reciprocal inhibitory effect on the antagonist motoneuron pool appears only in connection with the onset of the impulse activity of the agonist motoneuron pool--evidence that this effect is apparently the result of the activity of the spinal apparatus of reciprocal inhibition. As will be shown later, the organization of voluntary movement involves a supraspinal "tuning" of this apparatus (see Section 3 below). This apparatus may provide particularly for the reciprocal interaction of the motoneuron pools of antagonist muscles during the performance of a voluntary movement.

We must discuss briefly a seemingly paradoxical situation that was observed in the course of our investigations. On the one hand, rapid flexion of the foot is accompanied by the "concomitant" impulse activity of the motoneuron pool of the antagonist Sol; on the other hand, monosynaptic testing reveals a decline in the reflex excitability of this impulse-active antagonist motoneuron pool below the resting value for the "silent" pool. The presence of "concomitant" antagonist activity during the performance of voluntary movements has been observed by many investigators, and has led some to reject the applicability of the law of reciprocal inhibition to natural voluntary movements (Person, 1965).

However, reflex excitability testing of the antagonist motoneuron pool has shown that by itself, the presence of its "concomitant" impulse activity does not mean the absence of reciprocal inhibitory interaction between the motoneuron pools of the antagonist muscles. In this respect, our findings are in line with those of other investigators who have employed the reflex method in testing the excitability of the antagonist motoneuron pool (Hoffmann, P., 1922; Wachholder and Altenburger, 1927; Paillard, 1959; Mayer and Mawdsley, 1965; Baranov-Krylov, 1969; Gottlieb and Agarwall, 1972). Two explanations may be offered to account for this apparent paradox.

The first explanation is that the decline in the test H-reflex of the antagonist is the result of presynaptic inhibition. In accordance with this hypothesis, it must be assumed that the organization of voluntary movement includes the supraspinally evoked presynaptic inhibition (afferent depolarization) of the antagonist H-reflex, which leads to the diminution of the test H-reflex. The "concomitant" impulse activity of the Sol (antagonist) motoneurons is ensured by facilitatory actions on the motoneuron pool of this muscle; these actions are transmitted through pathways other than the H-reflex afferents (see Chapter 3, Section 3). Such a hypothesis appears to be contradicted by experimental data obtained in acute experiments with cats (Carpenter et al., 1962a, 1963; Andersen et al., 1964; Kostyuk, 1973), whereby the stimulation of the motosensory cortex evokes the depolarization of various primary afferent fibers, but not of the I-a muscle afferents.

As was shown in Chapter 2, Section 2, however, the evocation of a test H-response from the Sol with an amplitude up to 40% of the maximal (which was the response used in the present experiments) is most likely associated with the stimulation of the lowest-threshold I-a afferents. It may be, of course, that the mechanisms of corticospinal control in man differ in this respect from those in the cat. Also, the effects of natural corticospinal impulsation may differ substantially from those observed as a result of electrical stimulation of the motor cortex in acute experiments. Moreover, it is quite probable that the control of voluntary movements involves the participation of a number of descending systems, and not just the corticospinal system (see the Conclusion). It is known that besides the cortico- and rubrospinal tracts, other descending tracts in the cat can evoke the depolarization of the terminals of all the primary afferents, including the I-a afferents (see Chapter 3, Section 4).

According to the second explanation, the decline in the test H-response of the antagonist Sol accompanying the voluntary impulse activity of its motoneuron pool may be due to the fact that the motoneurons activated by the test stimulation of the low-threshold H-reflex afferents are comprised essentially of motoneurons that are reciprocally inhibited, while many of the impulse-active motoneurons are not excited by the testing afferent volley. The results of studies that support this explanation are given in the next section.

2. "OVERLAPPING" OF THE MOTONEURON POOLS OF ANTAGONIST MUSCLES

In an effort to explain one of the mechanisms of the origin of the "concomitant activity" of the antagonist during the performance of voluntary movements, we shall present our observations on the presence of a "concomitant" reflex response during stimulation of the antagonist nerve. Such "concomitant" responses are recorded in particular in the ATM during evocation of the "classic" H-response of the GM (Kots, 1969) and in the GM during evocation of the H-response of the ATM (Zhukov, 1971).

"Concomitant" H-Response of the ATM

While describing the characteristic features of the reflex H-response, P. Hoffmann (1922) and Magladery et al. (1952) noted that it is recorded in the foot extensors, but is absent in the foot flexors, particularly in the ATM. According to Magladery and co-workers, normally the late reflex response (F-response, according to the classification used by the authors) is recorded over the ATM during very intense stimulation of the PN, sufficient to evoke a strong M-response from this muscle. In the case of compression of the spinal cord, however (in the cervical and thoracic regions), the reflex response of the ATM is elicited at stimulus strengths that are subthreshold for the M-response. An increase in stimulus strength evokes an M-response from the ATM, and during a strong M-response, the reflex response of the ATM is diminished. According to data obtained by French et al. (1961), no H-responses are elicited from the ATM in healthy subjects, but are observed in patients suffering from phenylketonuria and congenital hemiplegia. Hohmann and Goodgold (1961) observed an H-response from the ATM in only 2 of 35 healthy subjects (5%), but in 75% of patients with lesions of the "upper motoneurons" of the brain.

We now know that all the studies mentioned above made use of inadequate amplification in recording the H-response of the ATM. At sufficient amplification, however, a "concomitant" H-response is recorded from the ATM together with the H-response of the GM in 100% of healthy subjects. We shall use the term "concomitant" in describing the response of the ATM that appears simultaneously ($\pm$2.5 msec) with the "classic" H-response of the GM (see Fig. 1). The principal contrast between the "concomitant" H-response of the ATM and the H-response of the GM lies in the much smaller amplitude of the former. For example, the amplitude of the H-responses of the GM is on the order of several millivolts; that of the "concomitant" H-response of the ATM, only a few tenths of a millivolt. The H-response of the GM at rest amounts to more than 50% of the maximal H-response (see Chapter 2, Section 1); that of the ATM, no more than 5%. The amplitude of the maximal H-response of the GM is always greater than even the strongest oscillations in the interference EMG of this muscle (see Fig. 20, for example), while the "concomitant" H-response of the ATM is appreciably weaker than the high-amplitude oscillations in the interference EMG that are characteristic of this muscle.

One circumstance makes it difficult to judge the nature of the "concomitant" H-response of the ATM--the fact that this response changes just as does the H-response of the GM in a variety of situations. In our investigations, such parallelism was observed in the following situations: at rest during an increase in stimulus strength (see Fig. 1); prior to voluntary extension and flexion of the foot (see Fig. 20); during electrically evoked vestibulospinal facilitation (see Chapter 5, Section 1); during ischemic deafferentation (see Figs. 6B and C); and during PTP (see Figs. 7A and D). The T-responses of the GM and the ATM also undergo parallel changes (Figs. 24A and B). These facts make it doubtful that the H-response of the ATM is associated with the reflex discharge of the MU of the ATM itself, and suggest that this response is the result of an influx of a strong H-response from the nearby heads of the GM. If this "influx" hypothesis is

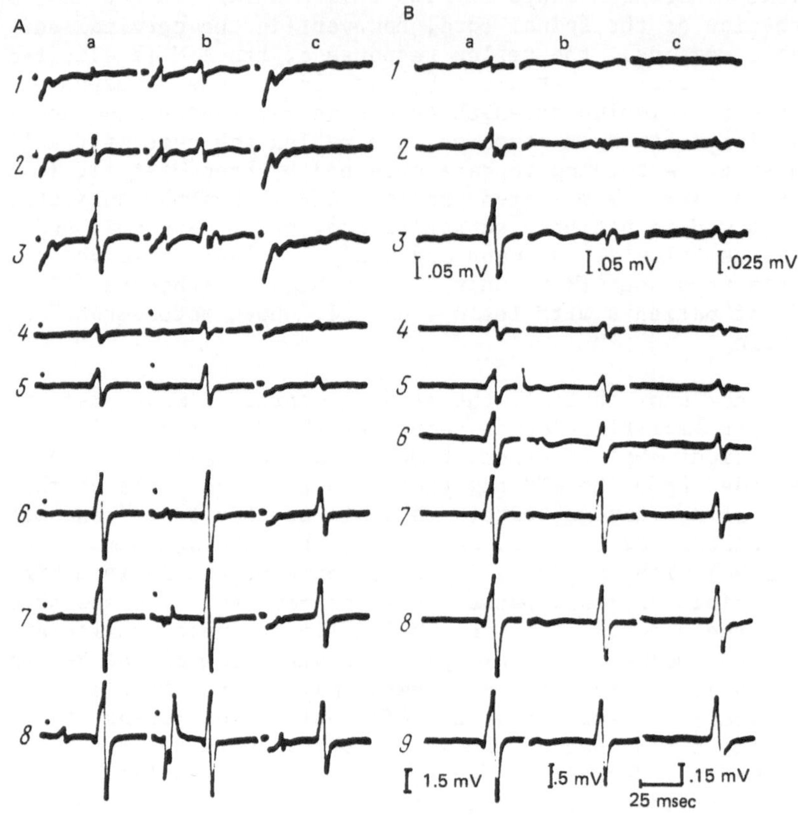
A
a
b
c
1
2
3
4
5
6
7
8
B
a
b
c
1
2
3
.05 mV
.05 mV
.025 mV
4
5
6
7
8
9
1.5 mV
.5 mV
.15 mV
25 msec

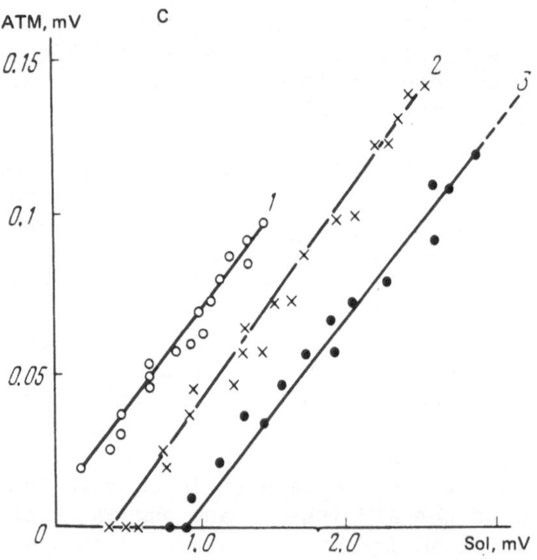
ATM, mV
C
0.15
0.1
0.05
0
1.0
2.0
Sol, mV
1
2
3

correct, the ratio of the amplitudes of the reflex responses of the GM and ATM should be the same in all cases. That this is not true, however, is shown by the following facts:

1. With a more lateral position of the stimulating electrode in the popliteal fossa, the reflex H-response of the ATM has a lower threshold than the H-response of the Sol, while with a more medial placement, the H-response of the Sol has the lower threshold (Figs. 25A and B).

2. The maximal "concomitant" H-response of the ATM evoked by a more medially positioned electrode is somewhat stronger than the maximal "concomitant" H-response of the ATM evoked by a more laterally positioned electrode (cf. traces C, 3, and D, 4, in Fig. 1 and traces A, 7, and B, 6, in Fig. 25). At the same time, the simultaneously recorded H-response of the Sol is stronger for a lateral than for a medial stimulating electrode (cf. traces A, 3, and B, 5, in Fig. 1; traces C, 3, and D, 4, in the same figure; and traces A, 7, and B, 6, in Fig. 25).

3. During supramaximal stimulation, the degree of depression of the ATM and Sol H-responses may be different for different positions of the stimulating electrodes. Thus, records are shown in Fig. 1 with equal H-responses of the Sol but different H-responses of the ATM (cf. traces C, 4, and D, 6).

4. The ratio of the amplitudes of the GM and ATM reflex responses is different during an electrically evoked H-reflex and a mechanically evoked T-reflex. Figure 24C shows graphically the results of one experiment in which the responses of the Sol, MG, and ATM were recorded simultaneously during the electrical stimulation of the popliteal nerve (medial position), a blow to the Achilles tendon (Achilles reflex), and a blow to the tendon or the ATM itself (tibial reflex), while the position of the recording electrodes and the amplification factor in all channels were kept constant. It can be seen that the slope of the line relating the amplitude of the reflex response of the ATM to the amplitude

---

Fig. 24. H- (A) and T-responses (B) of Sol (a), MG (b), and ATM (c). (A,B) (1-3) Threshold responses at high amplification (see calibration below record B, 3); (4-9) with increasing stimulus strength to nerve (A) or blow to Achilles tendon (B). (A, 8, and B, 6) Maximal reflex responses at rest; (B, 7-9) T-responses 1 min after removal of tourniquet. (C) Relationship between amplitude of reflex response of ATM (ordinate) and Sol (abscissa) on evocation of tibial reflex (1), Achilles reflex (2), and H-reflex (3).

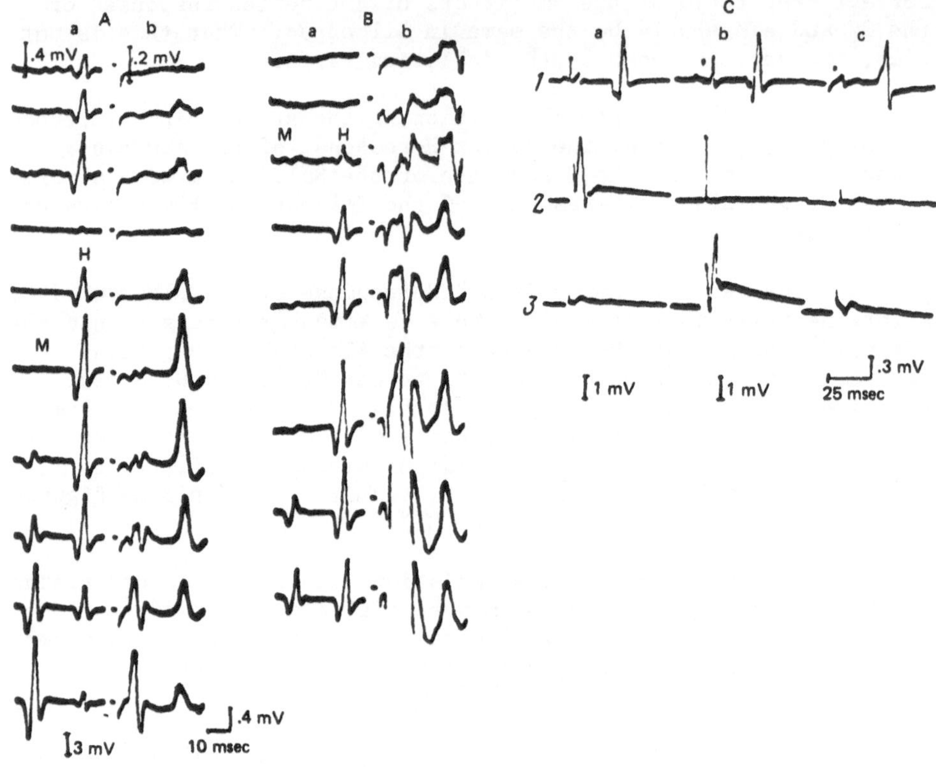

Fig. 25. H- and M-responses of Sol (a) and ATM (b) for medial (A) and lateral (B) positions of the stimulating electrode. (A,B) Strength of nerve stimulus increases from top to bottom. Top three traces recorded at higher amplification than other traces. (C) EMG of MG (a), LGM (b), and ATM (c) on evocation of H-reflex (1), and on direct stimulation of MG (2) and LGM (3). Further explanation in text.

of the reflex response of the Sol is more or less the same in all three cases, but the position of the lines in the diagram varies (which is also true of the ratio ATM/MG).

All these facts do not yet allow us to completely discount the role of electrical "influx" in the origin of the "concomitant" ATM responses, but they definitely show that the latter are--at least in part--proper responses of the ATM associated with the reflex excitation of its MU.

For a more direct determination of the possible role of physical "influx" in the origin of the ATM response accompanying the H-response of the GM, we performed special experiments in which the amplitude of the ATM response that could be recorded on evocation of the maximal H-response of the GM during nerve stimulation was compared with that possible on direct electrical stimulation of the GM, while the amplification factor was kept constant in all channels. In the latter case, we chose a stimulus that was adequate to evoke an M-response from the GM of approximately the same amplitude as the maximal H-response of this muscle. In the case of direct stimulation of the GM, a simultaneous "concomitant" ATM response can arise only from "influx." As can be seen in the records of Fig. 25C, strong M-responses of the Sol, MG, or LGM (m. gastroconemius lat.) evoked by direct stimulation with an amplitude equal to or even greater than the amplitude of the maximal H-response of the GM are accompanied by no M-response of the ATM, or a very small response. In the latter case, it is many times smaller than the amplitude of the ATM H-responses that are recorded simultaneously with the maximal H-responses of the GM. These findings suggest that physical "influx" plays a very significant role in the origin of the "concomitant" H-response of the ATM.

Hence, the "concomitant" H-response of the ATM (as well as the reflex response of the ATM evoked by a blow to the Achilles and its own tendon) is associated primarily with the reflex discharge of the ATM motoneurons. Judging by the similarity of the latency of the ATM H-response, as well as by the character of the changes during ischemic "deafferentation" and the effects of PTP (see Fig. 7), the "concomitant" H-reflex of the ATM (like the H-reflex of the GM) is probably a monosynaptic reflex, the arc of which includes low-threshold afferent fibers. As far as can be determined by its amplitude, the "concomitant" H-response of the ATM is evoked by the reflex discharge of relatively few of the motoneurons of this muscle.

It is quite significant that the "concomitant" H-reflex arc of the ATM apparently includes afferent fibers from the antagonist GM. Such a conclusion is based on the following observations: As was mentioned earlier, with a more lateral position of the stimulating electrode, the threshold for the M-response of the ATM is lower than for the "concomitant" H-response, in which case the H-response of the ATM follows a strong M-response (see Figs. 1 and 25A and B). If we consider that the "concomitant" H-response of the ATM is evoked by the stimulation of its own afferents, we might well conclude that the lowest-threshold motor axons of the ATM have a lower threshold than the group of ATM afferents the excitation of which is required for the reflex activation of the ATM motoneurons during evocation of the "concomitant" H-response of the ATM. On the other hand, this appears to be contradicted by the fact that with a more medial position of the stimulating electrode, the "concomitant" H-response of the ATM can be evoked without a preceding M-response (see Fig. 1C), and even the maximal "concomitant" H-response of the ATM is preceded in this case by only a very weak M-response of the ATM. Moreover, if the "concomitant" H-response of the ATM were a function only of the activation of the afferent fibers of the ATM comprising the peroneal (and correspondingly the sciatic) nerve, a more or less equal ratio should be observed between the thresholds and amplitudes of the M- and H-responses of the ATM, regardless of the position of the stimulating electrode. As we can see, however, this is not the case. It should be added that the strongest "concomitant" ATM H-response is evoked with a more medial, rather than lateral, position of the stimulating electrode.

The most adequate explanation for all these facts appears to lie in the recognition that the "concomitant" H-response of the ATM is dependent to a certain degree on the activation of the low-threshold afferent fibers comprising the antagonist TN. In other words, it may be assumed that the low-threshold afferent fibers can have a reflex monosynaptic facilitatory effect on the motoneuron pool of the antagonist. Such a conclusion is in agreement with experimental data regarding the presence in the cat of monosynaptic excitatory connections between the low-threshold I-a muscle afferents and the motoneuron pools of the antagonists (Eccles, J. C., et al., 1957; Willis et al., 1966). Such connections in particular have been traced by Eccles and his colleagues for the I-a muscle afferents of the antagonist muscles of the ankle joint--the GM and ATM.

It is interesting to note in this regard that approximately one-third of all primary afferent collaterals in the lumbosacral region of the spinal cord form direct connections with the motoneurons of the antagonists. In particular, afferent fibers from the GM terminate on the motoneurons of the ATM; and afferents from the ATM, on the motoneurons of the GM (Scheibel and Scheibel, 1969).

According to the findings of Schlegel and Sontag (1970), in the cat, there are also crossed reflex facilitatory effects on the fusimotor (gamma-) motoneurons of the antagonist. In their experiments, tetanic stimulation of the low-threshold GM afferents evoked the reflex activation of the fusimotor (and sometimes the alpha-) motoneurons of the ATM and the EDL (extensor digitorum longus).

"Concomitant" Tendon (T-) Response of the Antagonist

It has been known for some time that on evocation of a tendon reflex, action potentials are observed in the antagonists of healthy subjects (at times) and particularly of patients with pyramidal pathology (Lewy, 1923; Wachholder and Altenburger, 1924; Foerster and Altenburger, 1933). Peimer and Perli (1950) also found some activity in biceps femoris on evocation of the patellar reflex in healthy adults and children (see also Babkin, 1967). According to data obtained by Popov (1972), evocation of the Achilles reflex in children is accompanied by electrical responses from both antagonists--the Sol and ATM--and evocation of the patellar reflex, by responses from the quadriceps and biceps femoris muscles. Analogous observations were made by Gurfinkel' and Pal'tsev (1972) in healthy adult subjects; according to the findings, the amplitude of the "concomitant" (by our terminology) electrical response of the antagonist on evocation of a tendon reflex (Achilles or tibial) is about 10% of the maximal M-response of the given muscle. An analogous phenomenon in dogs was described earlier by Nesmeyanova (1971).

Our experimental design was based on the following considerations: During electrical stimulation of the TN, stimulation of the peroneal (and the sciatic) nerve (see Chapter 2, Section 1) and, with it, the "concomitant" excitation of the ATM afferents, cannot be entirely avoided. On evocation of the Achilles reflex, the receptors of the GM are subjected to mechanical stimulation, but those of the ATM are not. As experiments have shown, however, in the latter case, a reflex response is recorded from the ATM as well as from the Sol (see Fig. 24). The ratio of the amplitude of the T-response of the ATM to that of the Sol during evocation of the Achilles reflex is even greater than the ratio of the H-response of the ATM to the amplitude of the H-response of the Sol (Fig. 24C). In other words, contrary to expectations, the evocation of the "concomitant" Achilles T-reflex of the ATM is facilitated even somewhat more than the evocation of the "concomitant" H-response of the ATM. It could be concluded, therefore, that the mechanical excitation of the receptors of the GM leads both to the activation of the agonist motoneurons and to the reflex activation of the antagonist motoneuron (of the ATM).

One cannot rule out, however, the possibility that a blow to the Achilles tendon also evokes the excitation of the ATM receptors, which in turn leads to the "concomitant" reflex T-response of the ATM. If this is indeed the case, then a blow to the ATM or its tendon should elicit a T-response from the ATM with a maximum amplitude at least equal to (if not greater than) that evoked by a blow to the Achilles tendon. The maximum amplitude of the T-response of the ATM to a blow to its tendon or to the muscle itself (the tibial reflex) has proved to be smaller in most cases than the amplitude of the maximal T-response of the ATM to a blow to the Achilles tendon. At the same time, on evocation of the tibial reflex, the ratio of the amplitude of the ATM T-response to the amplitude of the GM T-response is greater than it is on evocation of the Achilles reflex, and particularly the H-reflex. Thus, stimulation of the ATM's own receptors facilitates evocation of the tibial T-reflex of this muscle, but is inadequate by itself to elicit a reflex response of the same magnitude as that evoked by the Achilles reflex. Consequently, it appears likely that afferentation from the GM (antagonist) as well as the possible afferentation from the ATM itself are involved in the reflex activation of ATM motoneurons during evocation of the Achilles reflex.

Thus, afferent input from the antagonistic GM evidently plays a certain role during both the electrically and mechanically evoked reflex activation of the ATM motoneuron pool. It is possible that a symmetrical, though not necessarily quantitatively equivalent, situation obtains for the facilitatory connections between the low-threshold afferents from the flexors (in particular the ATM) and the motoneuron pool of the GM extensors. This would explain in part why the largest GM H-response is evoked by a more lateral position of the stimulating electrode, i.e., close to the PN, just as the largest ATM H-response is evoked by a more medial position of the stimulating electrode, i.e., near the TN. In both cases, there is probably an amplification of afferent facilitatory input due to the concomitant excitation of the antagonist nerve afferents, without an amplification of antidromic effects. The experiments described in the next section will show that the low-threshold afferent fibers from the flexors (in particular the ATM) may indeed have monosynaptic excitatory connections with the motoneurons of the antagonist muscles (in particular with motoneurons of the GM).

"Intrinsic" ("Classic") H-Response of the ATM and "Concomitant" H-Response of the Sol

During stimulation of the PN (at the most lateral margin of the popliteal fossa at the tibia, approximately at the level of the fibular head), the polyphasic M-response of the ATM usually has the lowest threshold, and increases with increasing stimulus strength (Fig. 26B). The weak reflex response of the ATM (F-response, according to Magladery et al., 1952) appears only at stimulus strengths sufficient to evoke a substantial M-response; in this case, a very weak M-response is recorded from the Sol.

In some subjects, however, it is possible to elicit an "intrinsic" H-response from the ATM by stimulation of the PN (Fig. 26C). We shall use this term to designate the electrical response of the ATM that is evoked by a stimulus to its own (peroneal) nerve, thereby distinguishing it from the "concomitant" H-response of the ATM that is evoked by stimulation of the TN or the sciatic nerve or both. The prime distinction between these two reflex responses of the ATM is that evocation of the "concomitant" H-response of the ATM is accompanied by a strong H-response from the GM (see Fig. 26A), while evocation of the "intrinsic" H-response of the ATM is accompanied by very weak "concomitant" H-responses from the GM (see Fig. 26C).

The "intrinsic" H-response of the ATM has the following characteristics that make it entirely analogous to Hoffmann's "classic" GM H-response (Ya. M. Kots and V. I. Zhukov): It has the standard latency when evoked repetitively, equivalent to the latency (accuracy of measurement ±2.5 msec) of the H-response of the GM. The threshold for the "intrinsic" H-response of the ATM is occasionally lower than that for the M-response of this muscle, but in every case during weak suprathreshold stimulation of the PN, its amplitude is greater and it increases more rapidly with increased stimulus strength than the amplitude of the M-response (see first 5 traces from top to bottom in Fig. 26C, b). The amplitude of the "intrinsic" H-response of the ATM decreases as the M-response grows stronger (see traces 6-8). An H-response of the ATM with the same characteristics was described independently of us by Baranov-Krylov (1969).

Both the M- and H-responses of the ATM are characterized by the presence of several phases, possibly suggesting the incorporation of fibers with a wide range of conduction velocities into the PN. The nature of the later phases of the M-response, however, merits special study.

During the latent period and at the onset of voluntary flexion or extension of the foot--movements in which the ATM assumes the respective roles of agonist and antagonist--the "intrinsic" H-response of the ATM undergoes changes analogous to those observed during testing of the latent period of foot extension and flexion--movements in which the Sol plays the respective roles of agonist and antagonist--by the H-response of the Sol. Figure 26 shows fragments of records made in one experiment in which the "intrinsic" H-response of the ATM was utilized in the reflex excitability testing of the motoneuron pool of this muscle during the latent period and at the onset of flexion (Fig. 26D) and extension of the foot (Fig. 26E). In the first of these movements, the ATM assumes the role of agonist; in the second, that of antagonist. The average results from 8 experiments (3 subjects) involving foot flexion and 4 experiments (3 subjects) involving foot extension are given in Table 8. In both cases, the amplitude of the H-response is expressed as a percentage of the average amplitude throughout the first half of the latent period (>60 msec).

Our data indicate that during the last 60 msec of the latent period preceding the onset of flexion, the "intrinsic" H-response of the ATM (agonist of the future movement) increases steadily as the onset of movement draws near. At the onset of the voluntary

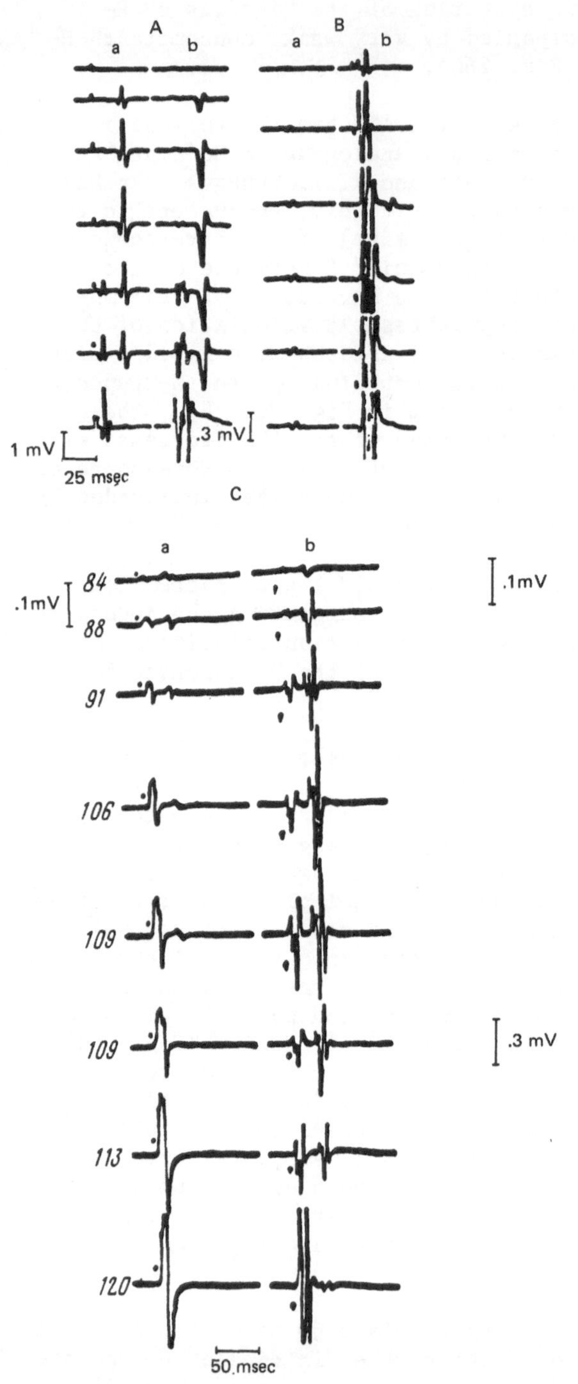
A
a
b
B
a
b
1 mV
25 msec
.3 mV
C
a
b
84
88
91
106
109
109
113
120
.1mV
.1mV
.3 mV
50 msec

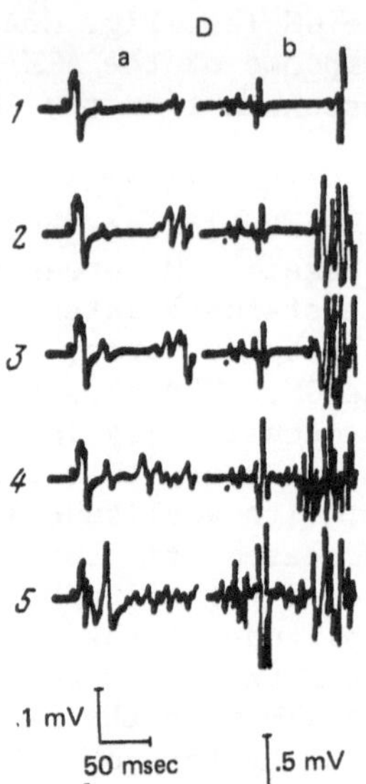
D
a
b
1
2
3
4
5
.1 mV
50 msec
.5 mV

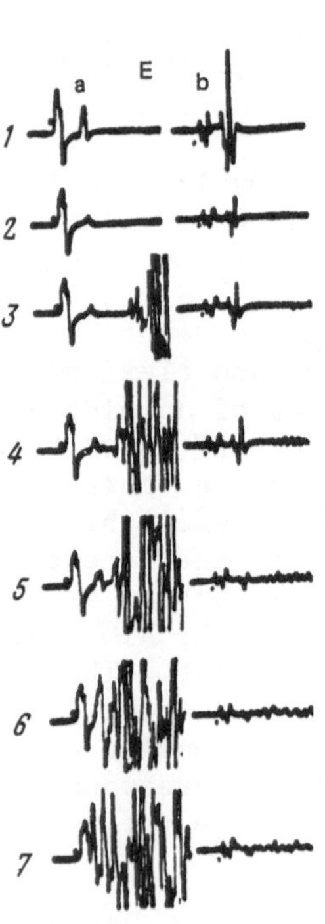
a
E
b
1
2
3
4
5
6
7

EMG volley of the ATM, the amplitude of the test "intrinsic" H-response of this muscle is several times greater than under control conditions. During the entire latent period (±20 msec) preceding the onset of foot extension, the "intrinsic" H-response of the ATM (the future antagonist of the voluntary movement) remains unchanged. At the onset of the voluntary EMG volley of the GM (agonist), the amplitude of the test "intrinsic" H-response of the ATM (antagonist) is sharply reduced.

Thus, judging by the basic functional signs, the "intrinsic" H-response of the ATM recorded in some subjects as evoked by stimulation of the PN is similar to the H-response of the GM evoked by preferential stimulation of the TN. Like the H-response of the GM, the "intrinsic" H-response of the ATM increases during the last 60 msec of the latent period of a voluntary movement (as well as during the movement) in which this muscle serves as the agonist. This reflex response remains unchanged during the course of the latent period (±20 msec) and is sharply depressed at the onset of a voluntary movement in which the ATM serves as antagonist. These facts allow us to conclude that the "intrinsic" H-response of the ATM elicited by stimulation of the PN is a result of the reflex activation of the motoneurons of this muscle, and is analogous by nature to Hoffmann's "classic" ("intrinsic") H-response of the GM.

As was mentioned earlier, a "concomitant" H-response is recorded from the GM along with the "intrinsic" H-response of the ATM (see Figs. 26C-E). This term "concomitant" H-response may be used in designating the weak reflex responses of the GM that appear during stimulation of the PN and behave in like manner as

---

Fig. 26. "Intrinsic" and "concomitant" H-response at rest (A-C) and prior to voluntary movement (D,E). (A) "Intrinsic" H-response of Sol (a) and "concomitant" H-response of ATM (b) during stimulation of TN; (B) M- and H-responses of Sol (a) and ATM (b) during stimulation of PN; (C) "intrinsic" H-response of ATM (b) and "concomitant" H-response of Sol (a) during stimulation of PN. (A-C, top to bottom) Stimulus strength to nerve increases. Last three traces under C: responses of ATM at lower amplification than other traces. (D,E) "Intrinsic" H-response of ATM (b) and "concomitant" H-response of Sol (a) during latent period and at onset of voluntary flexion (D) and extension (E) of foot. (D) H-responses prior to (1-3), accompanying onset of (4), and subsequent to onset of (5) voluntary EMG volley of ATM; (E) maximal (1) and control H-responses of ATM at rest (2), prior to (3-5), and during (6,7) voluntary EMG volley of GM.

Table 8. Average Amplitudes of "Intrinsic" H-Response of ATM During Latent Period and at Onset of Flexion and Extension of Foot

	Interval from H-response of ATM to EMG volley of primary agonist (msec)					Interval from onset of EMG to H-response of ATM (msec)	
Index	60	50	40	30	20	10-30	40-60
Flexion							
M	100	120.5	157.4	207.7	304.6	399.3	443.7
±m	--	±3.2	±12.8	±23.8	±46.4	±53.6	±62.3
min-max	--	110-131	113-225	126-360	129-484	250-600	300-675
N	8	8	8	8	8	7	8
n	22.4	11.1	11.2	12.9	12.9	11.0	12.3
Extension							
M	100	100.7	98.9	98.5	60.0	35.0	26.5
±m	--	±1.0	±3.8	±3.7	±6.3	±5.6	±3.2
min-max	--	99-103	94-108	94-108	33-88	25-51	20-33
N	4	4	4	4	4	4	4
n	22.5	10.3	11.0	11.5	11.8	12.0	11.8

the "intrinsic" H-response of the ATM during the course of the latent period. Indeed, as the records in Fig. 26D indicate, the gradual increase of the "concomitant" H-response of the Sol (the future antagonist of the movement) during the last 60 msec of the latent period of voluntary flexion parallels that of the "intrinsic" H-response of the ATM (the future agonist of the movement). The same "concomitant" H-response of the Sol, like the "intrinsic" H-response of the ATM, exhibits no noticeable changes during the entire latent period of foot extension--a movement in which the Sol must serve as antagonist (see Fig. 26E). The changes in the "concomitant" H-response of the Sol during foot extension (like the changes in the "concomitant" H-response of the ATM during foot flexion) cannot be traced, because the response is masked by the strong oscillations of the voluntary EMG volley of the agonist.

We shall now summarize the principal results of our study on the "concomitant" H-response of the antagonist muscles of the ankle joint--the ATM and GM. During the preferential stimulation of one of the two nerves connected to these muscles (the TN or the PN), besides the strong "intrinsic" H-response of the muscle innervated by the stimulated nerve, a weak "concomitant" reflex H-response is evoked from the antagonist. The latter response is most likely the result (at least for the most part) of the reflex firing of a small number of motoneurons having monosynaptic facilitatory connections with the low-threshold H-reflex afferents from the antagonist muscles. If this is actually the case, then we must conclude that the motoneuron pools of the antagonist muscles of the ankle joint in man have more or less common sources of facilitatory reflex influences. In particular, the low-threshold H-reflex afferents establish monosynaptic excitatory contacts with the motoneurons of the antagonist muscles, as well as with their own motoneurons.

It is quite significant that during the course of the latent period, the state of the motoneurons that are reflexly activated by preferential stimulation of the afferent fibers of the antagonist nerve changes in the same manner as the reflex excitability of the antagonist motoneuron pool. That it does suggests that besides common sources of identical reflex effects, some motoneurons of the antagonist muscles have common sources of identical supraspinal effects. In this sense, there is apparently a partial "overlapping" of the motoneuron pool of the antagonist muscles of the ankle joint in man.

A functional "overlapping" of this sort may account particularly well for the presence of "concomitant" impulse activity that is almost always observed in the motoneuron pool of the antagonist during the performance of natural movements. The supraspinal (in origin) facilitation of the antagonist motoneurons from the "overlap" zone is revealed by the increase in the amplitude of the test "concomitant" H-response of the antagonist during the course of the latent period and at the onset of movement (see Fig. 20A, H-response of ATM prior to foot extension; Fig. 26D, H-response of Sol prior to foot flexion). It is quite probable that these motoneurons are also impulse-active during the phasic voluntary movement, which determines the presence of the "concomitant" activity of the antagonist (antagonist EMG).

It should be noted once more that the "intrinsic" and "concomitant" H-responses of the same muscle behave differently during the course of the latent period and at the onset of voluntary movement. For example, the "intrinsic" H-response of the Sol increases, while the "concomitant" H-response of the Sol does not change, during the latent period of foot extension. The "intrinsic" H-response of the Sol does not change during the latent period and declines sharply at the onset of foot flexion, while the "concomitant" H-response of the Sol increases during the course of the latent period of this movement. The behavior of the "intrinsic" and "concomitant" H-responses of the ATM is analogous during the course of the latent period and at the onset of flexion and extension of the foot.

This difference in the behavior of the "intrinsic" and "concomitant" H-responses of the same muscle may be understood by assuming that these two types of muscle responses are associated with the reflex discharge of essentially (if not entirely) different motoneurons. In particular, this would suggest that the state of motoneurons from the "overlap" zone cannot be determined during testing by the "intrinsic" ("classic") H-reflex. We can then understand the "paradoxical" situation described above involving the depression of the test "intrinsic" H-response of the antagonist against the background of some "concomitant" impulse activity by its motoneuron pool. The "concomitant" impulse activity of the antagonist motoneuron pool may be essentially determined by the activity of that fraction of the motoneurons from the "overlap" zone the state of which is not tested by the "intrinsic" H-reflex of this pool.

Testing by the "intrinsic" H-reflex of the ATM during the latent period and at the onset of voluntary movements in which this muscle serves as agonist or antagonist (foot flexion or extension, respectively) has shown that the changes in the reflex excitability of the flexor motoneuron pool completely coincide with the changes (described above) in the state of the extensor motoneuron pool during the organization of movements in which the extensors assume the role of the agonists or the antagonists (extension or flexion, respectively). Analogous changes during the latent period of voluntary movement were observed in our laboratory and during testing by the H-reflex evoked in various muscles of the hand (Mart'yanov and Kopylov, 1971). It may thus be concluded that the supraspinal organization of different voluntary movements makes use of similar spinal mechanisms, regardless whether the flexors or extensors of the lower or upper limbs assume the role of agonists or antagonists in these movements.

3. THE INTERNEURONAL SPINAL APPARATUS OF RECIPROCAL ANTAGONIST INHIBITION OF THE ANTAGONIST

Electrophysiological studies of the last three decades involving the use of monosynaptic testing methods, intracellular recording, and several other methods have made it possible to obtain much information on the structure of the neuronal connections in the spinal cord that provide for the reciprocal interactions of antagonist muscles, as well as on the nature of reciprocal inhibition (for surveys, see Granit, 1955; Kostyuk, 1959; Eccles, J. C., 1957, 1964, 1969; Ioseliani, 1970).

The first study in this area was performed by Lloyd (1941b, 1946), who showed, by experimenting with the paired stimulation of the nerves of antagonist muscles, that the conditioning stimulation of the I-a afferents of one nerve leads to a decline in the test monosynaptic reflex of the antagonist. Lloyd found that the inhibitory and excitatory actions of I-a fiber impulses on the motoneurons of antagonist muscles have the same central delays. Since the testing excitatory volley in the I-a afferents act monosynaptically on the motoneurons of its muscle, Lloyd concluded that the inhibition of antagonist motoneurons is also realized via the monosynaptic central pathway. Evidently, Lloyd assumed, the branches of the I-a fibers form direct inhibitory synapses with the motoneurons of the antagonist. For this reason, Lloyd called the reflex reciprocal inhibition of antagonist motoneurons "direct inhibition."

However, later studies by Renshaw (1942), Kostyuk (1959), and particularly Eccles (1957) showed that one inhibitory interneuron is incorporated into the pathway responsible for "direct" inhibition. The structural scheme now accepted for "direct" inhibition is as follows: Impulses in the I-a afferent fibers emanating from the muscle spindles excite the homonymous motoneurons (and the motoneurons of the synergists) and inhibit the motoneurons of the antagonists. This reciprocal inhibitory action is associated with the excitation of the segmental inhibitory interneurons that are located in the intermediate nucleus and are activated by impulsation of the I-a afferents from the antagonists. These interneurons evoke the postsynaptic inhibition of motoneurons (Eccles, J. C., 1969).

The reflex reciprocal interaction of the motoneuron pools of antagonist muscles is not only limited to "direct" inhibition, but also is always accomplished with the participation of spinal interneurons. The inhibition in the motoneurons of antagonists that is evoked by I-b afferents is mediated by interneurons located in the ipsilateral intermediate nucleus (Eccles, J. C., 1964, 1969). Impulse frequency in group II and III afferents, and in cutaneous and high-threshold joint afferents, evokes the polysynaptic facilitation of flexor motoneurons, and has a reciprocal inhibitory influence on extensor motoneurons (Eccles, R. M., and Lundberg, 1959a,b; Holmquist and Lundberg, 1961). The spinal interneurons that are monosynaptically activated by impulse frequency of the flexor reflex afferents are located in the intermediate area and in the base of the dorsal horn (Lundberg et al., 1962; Hongo et al., 1966).

The control of the spinal mechanisms governing the reciprocal interaction of the motoneuron pool of antagonist muscles may be accomplished over different segmental reflex pathways. For example, evocation of the crossed extensor reflex is accompanied by the facilitation of the reciprocal inhibitory interneurons linked with the flexor motoneurons (Lundberg, 1970; Hultborn, 1972). Besides reflex influences from the receptors, the enhancement or attenuation of the reciprocal interaction of antagonistic motoneuron pools may also involve the participation of recurrent effects associated with the impulse activity of motoneurons.

According to the data obtained in a study by Hultborn (1972), recurrent effects from the active motoneuron pool may depress reciprocal "direct" inhibition from antagonist muscles. The interneurons, with a convergence of monosynaptic excitation from the I-a afferents and inhibition from the motoneuron axon collaterals, are located in the ventral horn dorsomedial to the motor nuclei (Rexed's lamina VII). These "ventral" I-a interneurons differ from the I-a interneurons of the intermediate area, which were postulated by J. Eccles (1969) to be I-a reciprocal inhibitory interneurons, the difference being mainly that the latter are not subjected to recurrent inhibition and receive no monosynaptic excitatory actions from the vestibulospinal tract, although they are subject to descending facilitation from the pyramidal (Lundberg et al., 1962) and rubrospinal tracts (Hongo et al., 1965; Kostyuk, 1973), as are the "ventral" I-a interneurons.

Some data point to the possible participation of a presynaptic inhibitory mechanism in regulating the reciprocal interaction of the motoneuron pools of the antagonist muscles of various joints (Eccles, J. C., et al., 1962; Sverdlov, Yu. S., 1967; Barnes and Pompeiano, 1970). But which spinal mechanisms are involved in the supraspinally determined reciprocal interaction of the motoneuron pools of antagonist muscles?

One of the mechanisms of the supraspinal reciprocal influence on motoneurons appears to be the activation of interneurons of the segmental reflex arcs. In experiments with cats immobilized by

Flaxedil, Lundberg and Voorhoeve (1962) determined that the corticospinal influences that are mediated by the pyramidal tract enhance the reciprocal spinal effects from the I-a, I-b, flexor reflex, and cutaneous afferents by their excitatory action on the interneurons of the corresponding reflex arcs. As in spinal preparations (Eccles, R. M., and Lundberg, 1959a), in cats immobilized by Flaxedil, "direct" reciprocal inhibition (from the I-a flexor afferents to the extensor motoneurons of the ankle joint) is frequently absent or of very low prominence. On the other hand, it is strongly facilitated by cortical (pyramidal) stimulation. Lundberg and Voorhoeve consider it quite likely that, as with other spinal reflexes, the normal functioning of reflex reciprocal "direct" inhibition is dependent on the supraspinal (pyramidal) facilitation of the spinal interneurons incorporated into this reflex path.

This hypothesis was corroborated in another study by Lundberg et al. (1962) involving intracellular recording from interneurons of the dorsal horn and the intermediate nucleus. It was determined that stimulation of the sensorimotor cortex evokes EPSPs in many interneurons activated by the following types of primary afferents: I-a muscle afferents, flexor reflex afferents, and cutaneous afferents. According to Chambers and Liu (1958), these interneurons constitute the area of termination of endings from the pyramidal tract in the cat. In interneurons that are monosynaptically activated from group I muscle afferents, the appearance of EPSPs during cortical stimulation precedes the onset of the cortical amplification of "direct" reciprocal inhibition by at least 2-3 msec.

The participation of spinal interneurons in the supraspinal reciprocal inhibition of motoneurons has also been demonstrated in experiments with monkeys (Preston and Whitlock, 1960, 1961; Hern et al., 1962; Landgren et al., 1962a,b). Preston and Whitlock (1960) observed the discharge of an interpolated neuron during the period of time required for the transmission of the inhibitory influence to the motoneurons, and suggested that the higher sensitivity of the inhibitory pathway to the action of barbiturates is explained by the presence of the interpolated neuron. Activation of the inhibitory spinal interneurons is apparently accomplished via the collaterals of the same descending corticospinal fibers that form monosynaptic connections with spinal motoneurons. It is well known that no summation is required for the output effect of the inhibitory interneurons in their depressive action on motoneurons (Stewart and Preston, 1967). This property might also account for the ease with which the inhibition of impulse-active motoneurons is evoked by stimulation of the motor cortex (Sherrington, 1906; Gellhorn, 1956).

Other descending motor systems besides the pyramidal tract can have reciprocal effects on the motoneuron pools of antagonist muscles by acting on the spinal interneurons of reciprocal action.

As has been shown by Hongo et al. (1965, 1969b) and Lundberg (1966), stimulation of the red nucleus enhances the reciprocal inhibitory effect of the I-a and I-b afferents of antagonist muscles, as well as the facilitatory and inhibitory effects on the motoneurons of the low-threshold joint and skin afferents and the flexor reflex afferents. It is assumed that such effects are based on the facilitation of interneurons of the reflex arcs located particularly in the intermediate nucleus of the gray matter of the spinal cord (Shapovalov and Shapovalova, 1966; Kostyuk and Pilyavskii, 1967, 1969; Shapovalov and Karamyan, 1968; Hongo et al., 1969a,b). It has been shown in studies of recent years that the vestibulospinal tract has an influence on various interneuronal systems of the spinal cord (Erulkar et al., 1966), including a monosynaptic facilitatory influence on the interneurons of "direct" inhibition (Grillner et al., 1966; Bruggencate et al., 1969).

Supraspinal systems can also exert an inhibitory influence over the dorsal and ventral reticulospinal pathways, as well as over two monoaminergic pathways that, in all likelihood, originate in the brain stem (Lundberg, 1966). It was shown in a number of studies that stimulation of the reticular formation of the brain stem produces inhibition of the spontaneous and evoked activity of the spinal interneurons (Koizumi et al., 1959; Engberg et al., 1968) that are monosynaptically activated by peripheral afferents. These findings allow us to conclude that the influence of the reticulospinal system on the specialized interneuronal segmental apparatus may serve as one of the mechanisms of the reciprocal action of this system on the motoneuron pools of antagonist muscles.

It was determined in studies by Hultborn (1972) that the excitatory and inhibitory influences from various segmental and supraspinal origins that control the I-a reciprocal inhibitory effects converge at the same I-a inhibitory interneurons.

Another mechanism that determines (or enhances) the reciprocal effects evoked by supraspinal influences may be the supraspinal control of the fusimotor (gamma-) motoneurons (Matthews, P. B. C., 1964). A number of studies have demonstrated that the supraspinal influences on gamma-motoneurons have a marked reciprocal character, the effects paralleling those on the alpha-motoneurons: activation of the pyramidal system leads to the facilitation of flexor gamma-motoneurons and the inhibition of extensor gamma-motoneurons (Kato et al., 1964; Laursen and Wiesendanger, 1966; Fidone and Preston, 1969); stimulation of the Deiters' nucleus leads to the facilitation of extensor gamma-motoneurons (Carli et al., 1967; Pompeiano et al., 1967; Grillner et al., 1969); stimulation of the medial longitudinal fasciculus leads to effects similar to the effects of pyramidal stimulation (Grillner et al., 1969). It should be noted that the latent period of the reciprocal influences on gamma-motoneurons is shorter than that of those on alpha-motoneurons (Granit, 1955; Buchwald et al., 1961; Mortimer and Akert, 1961; Laursen and Wiesendanger, 1966; Koeze et al., 1968; Fidone and Preston, 1969).

Hongo et al. (1969b) believe that the alpha-gamma linkage may serve as the basic mechanism for the control of reciprocal inhibition. The concept of "alpha-gamma coupled reciprocal inhibition" is based on data indicating that a number of neuronal pathways that evoke parallel excitatory effects in alpha- and gamma-motoneurons of the agonists also evoke the excitation of the I-a inhibitory interneurons that act on the motoneurons of their antagonists. In this way, a volley in the corticospinal tract may evoke not only the coactivation of alpha- and gamma-motoneurons (Mortimer and Akert, 1961; Vedel, 1966; Fidone and Preston, 1969)--though such coactivation may be absent in the baboon (Koeze, 1968; Koeze et al., 1968)--but also the corresponding facilitation of transmission in the I-a inhibitory pathway (Lundberg and Voorhoeve, 1962). Rubrospinal-tract activity excites the alpha-motoneurons of the flexors and, to a certain extent, of the extensors, although the predominant effect for the extensor motoneurons is one of inhibition (Hongo et al., 1969a). The action of the rubrospinal pathway on gamma-motoneurons is still unclear, although this pathway has exhibited excitatory action on the "dynamic" gamma-motoneurons (Granit, 1970). Hongo et al. (1969b) have shown that the I-a inhibitory effect on flexors and extensors is facilitated via the rubrospinal pathways.

The vestibulospinal influences give the clearest example of parallel effects on the alpha- and gamma-motoneurons and the I-a inhibitory interneurons. The vestibulospinal tract evokes the monosynaptic excitation of the extensor alpha- and "static" gamma-motoneurons of the knee joint musculature, and of the I-a inhibitory interneurons that mediate reciprocal inhibition from the extensors to the flexor motoneurons of the knee-joint musculature (Grillner et al., 1970). The vestibulospinal reciprocal inhibition of the knee-joint flexors is effectively depressed by an antidromic volley in the ventral root, which suggests the mediation of this reciprocal inhibition mainly by the "ventral" I-a inhibitory interneurons (Hultborn, 1972).

The reticulospinal pathway in the medial longitudinal fasciculus has an effect on the lumbar motoneurons that opposes the effect of the vestibulospinal tract. Stimulation of the medial longitudinal fasciculus evokes the monosynaptic excitation of the alpha- and "static" gamma-motoneurons of the flexors, as well as the disynaptic inhibition of the extensors of the knee and ankle joints (Grillner et al., 1968). However, the latter effect is not subject to recurrent inhibition, while the facilitation of transmission in the I-a inhibitory pathway to these motoneurons is very weak, if present (Hultborn, 1972).

To the extent that the reciprocal interaction of antagonistic motoneuron pools is dependent on the state of the presynaptic inhibitory interneurons and the Renshaw cells, it may be modulated by the supraspinal control of these interneurons. Investigating the possible functional significance of the Renshaw cells in regulating the level of reciprocal inhibition, Hultborn (1972) was struck by the fact that the main recurrent inhibition of the I-a reciprocal inhibitory interneurons is evoked from the motor fibers of the same muscles the I-a afferents of which activate these interneurons. Hultborn proposed that the recurrent control of the I-a inhibitory pathway may have the function of maintaining a continuous reciprocal inhibition during different levels of agonist activity during the performance of movements dependent on the alpha-gamma linkage.

However, the performance of a variety of movements may require not only a constant degree of reciprocal inhibition during different degrees of agonist contraction, but also a mechanism for selecting the degree of inhibition itself. At times, it is necessary that reciprocal inhibition be entirely absent. Owing to the spinal and supraspinal control of the Renshaw cells, the transmission of impulses in the recurrent pathway may be enhanced or depressed, thereby regulating the degree of reciprocal inhibition. If the performance of a particular movement requires effective reciprocal inhibition, transmission in the recurrent pathway must be inhibited.

We have already mentioned (Chapter 3, Section 4) that the inhibition of the agonist Renshaw cells that precedes the onset of a voluntary phasic movement may be one of the mechanisms for achieving the reciprocal antagonist inhibition that is necessary for the performance of this type of movement. Conversely, during facilitation of the Renshaw cells, even weak activity by the agonist alpha-motoneurons will evoke the recurrent inhibition of I-a inhibitory interneurons, and will thus maintain the reciprocal I-a inhibition of the antagonist motoneurons at the very low level required for "nonreciprocal" movement (or static efforts).

Thus, experimental data obtained from acute experiments with animals allow us to assume that the reciprocal interaction of the motoneuron pools of antagonist muscles is achieved largely via the spinal interneuronal apparatus of reciprocal inhibition. It is quite significant that this spinal reciprocal inhibitory apparatus (SRIA) is evidently under the powerful control of the supraspinal motor centers.

These findings have given rise to the hypothesis that one of the fundamental mechanisms for the central control of the segmental centers of antagonistic muscles during the performance of a voluntary movement is the supraspinal regulation of the state of the SRIA, which in turn provides for the reciprocal interaction of the motoneuron pools of the antagonist muscles (Kots, 1968, 1969a, b). In particular, we believe that the organization of voluntary movement may include a supraspinal "tuning" of this spinal inhibitory apparatus in accordance with the character of the interaction of antagonist muscles required for the performance of a particular movement (Kots, 1969a,b; Kots and Zhukov, 1971). While this hypothesis was being developed, it was also postulated that the character of the supraspinal "tuning" of the SRIA may be one of the mechanisms that determine the great variety in the activity ratios of antagonistic motoneuron pools during the performance of different movements, as well as the changes in the activity ratios as a result of learning. We shall now present the results of experiments designed to test these hypotheses.

Method of Evaluating the State of the Spinal Reciprocal Inhibitory Apparatus (SRIA)

To test the state of the SRIA, we applied the method of Lloyd (1941) of paired stimulation--conditioning stimulation of the PN and test stimulation of the TN (Kots and Zhukov, 1971). The PN was stimulated with bipolar electrodes positioned near the outer edge of the popliteal fossa, lateral and distal to the electrodes for stimulating the TN. The electrodes were positioned over the PN in such a way that stimulation would evoke an M-response from the ATM with no significant "concomitant" M- and H-responses from the Sol (see Fig. 26B). The reflex response of the ATM recorded during stimulation of the PN usually has a small amplitude and a higher threshold than the M-response (see trace 3 in Fig. 26B). This reflex response declines as the M-response increases (see traces 3-6). Only occasionally can a low-threshold reflex ATM response be evoked without a preceding M-response (see the previous section).

In experiments with paired stimuli, the conditioning stimulus to the PN was chosen so as to evoke an ATM M-response with an amplitude approximately 30% of the maximal. During single-stimulus testing, a control Sol H-response was employed with an amplitude 20-40% of the maximal; during paired-stimulus testing, 50-60%. The amplitude of the test H-response of the Sol during paired stimulation was expressed as a percentage of the amplitude of the single control H-response.

The average results from 21 experiments (18 subjects) designed to determine the effect of simultaneous or prior conditioning stimulation of the PN on the amplitude of the test H-response of the Sol are shown graphically in Fig. 27C, 1; Fig. 27A shows the records of one of these experiments. As these data indicate, during simultaneous stimulation of the PN and the TN (zero delay), the amplitude of the test H-response of the Sol is substantially greater (by an average factor of 1.5) than the amplitude of the control H-response. If the conditioning stimulation of the PN precedes test stimulation of the TN by 1 msec, the increase in the amplitude of the test H-response over the control amplitude becomes significantly smaller than for zero delay. For interstimulus delays greater than or equal to 2 msec, the preceding stimulation of the PN produces a significant decrease in the reflex excitability of the motoneurons of the antagonist Sol below the background level.

As shown in the records in Fig. 27A, the M-responses of both the Sol (a) and the ATM (b) are larger for the simultaneous stimulation of both nerves (trace 3) than for a 1-msec delay (trace 4). But even in the latter case, the M-responses of the Sol and ATM are appreciably larger than for stimulation of the PN only (trace 2) (in this case, stimulation of the TN evokes no M-responses from the Sol and ATM--trace 1). During paired stimulation with a conditioning-testing interval of 2 msec and more (traces 5-11), the M-responses of each muscle are the same as during the single stimulation of the PN. These findings may stem from the fact that during stimulation of each of the nerves, the current loop apparently encompasses part of the other nerve. Hence, during the simultaneous stimulation of both nerves, the stimulus strength to each nerve is greater than during their individual stimulation. This is also true for an interstimulus delay of 1 msec to the two nerves, since the duration of the conditioning stimulation of the PN is equal to 1 msec; thus, for a 1-msec delay, the trailing edge of the conditioning stimulating impulse coincides with the leading edge of the testing impulse.

The influx of current to the other nerve may account for the increase in the amplitude of the H-response of the Sol when it is evoked by TN stimulation simultaneously with or 1 msec after stimulation of the PN. With such a delay, the electrical "addition" resulting from current influx during the stimulation of the PN, summed with the stimulation of the TN, results in a greater number of excited Sol H-reflex afferents than during single stimulation of the TN. There is an increase in the amplitude of the Sol H-response corresponding to the increased number of excited afferents.

Aside from its purely "artifactual" origin (from current influx), the increase in the test H-response of the Sol during the conditioning stimulation of the PN simultaneously with stimulation of the TN or with a 1-msec delay may be due to the fact that the conditioning stimulation of the PN is accompanied by the excitation

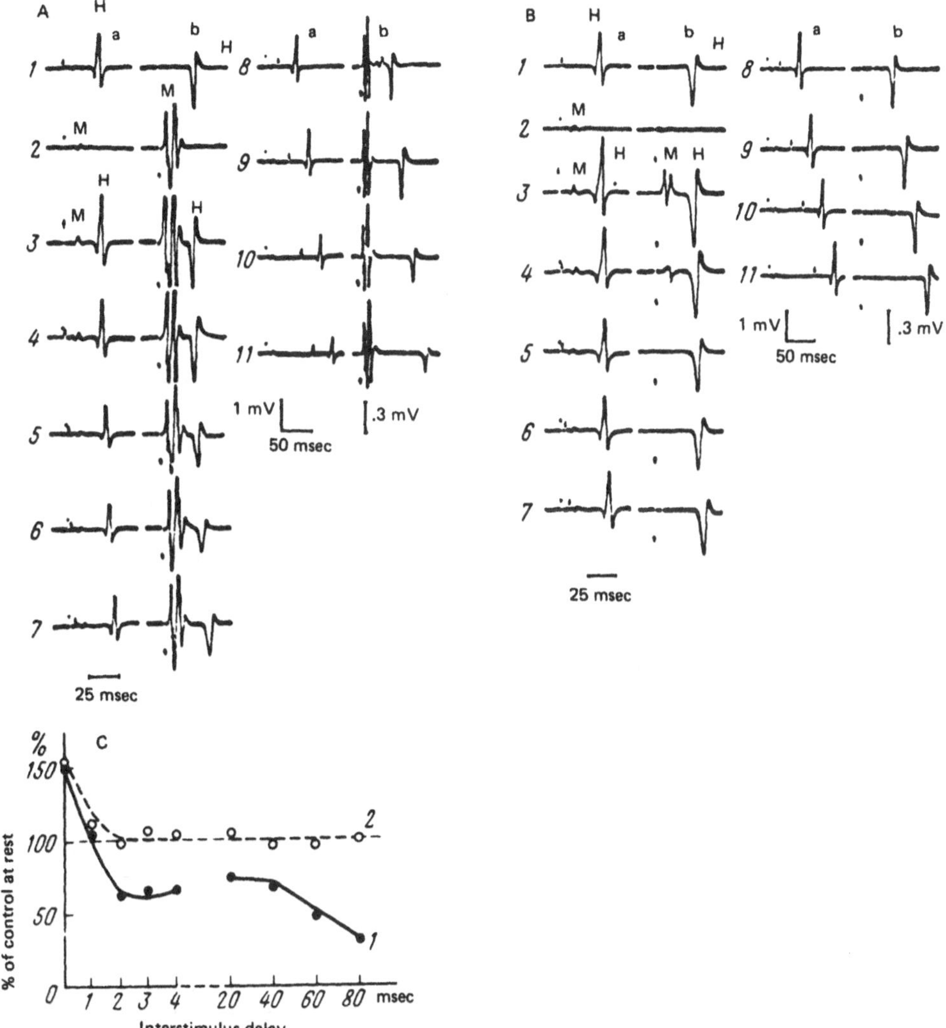

Fig. 27. Effect of conditioning stimulation of PN (A) and conditioning subthreshold stimulation of TN (B) on H-response of Sol at rest. (A,B) EMGs of Sol (a) and ATM (b); (1) single test stimulation; (2) single conditioning stimulation; (3-11) paired stimulation with zero delay (3) and with interstimulus delays of 1 (4), 2 (5), 3 (6), 6 (7), 20 (8), 40 (9), 60 (10), and 80 (11) msec. Traces 8-11 were made at lower beam sweep speeds than traces 1-7. (C) Amplitude of test H-response of Sol vs. delay between conditioning stimulation of PN (1) or subthreshold stimulation of TN (2) and test stimulation of TN.

of the afferent fibers of this nerve that may have monosynaptic facilitatory connections with the antagonist (Sol) motoneuron pool (see the previous section).

Inhibitory action is observed when the interval between the conditioning stimulation of the PN and the test stimulation of the TN is increased to 2 or more msec. The greatest inhibitory effect is observed for delays of 3-4 msec. It is noteworthy that during paired stimulation of the PN and the TN, a clear inhibitory effect is seen only in cases in which the conditioning stimulation of the PN evokes a strong M-response from the ATM (no less than 20-30% of maximal), i.e., when it evokes the excitation of a large number of PN fibers.

Since the conditioning stimulation of the PN is apparently accompanied by current flow to the TN, it was necessary to determine whether the subthreshold stimulation of the TN is one cause of the decline in the amplitude of the test H-response of the Sol during paired stimulation with a delay of 2 or more msec. In a special series of experiments, therefore, we compared the effects of the conditioning stimulation of the PN and the conditioning subthreshold stimulation of the TN. Three pairs of electrodes were attached in the area of the popliteal fossa--two pairs over the TN (one pair for conditioning subthreshold stimulation of TN, the other for test stimulation) and one pair over the PN.

The average results from 8 experiments of this series (5 subjects) are shown graphically in Fig. 27C, 2 (open circles). Figure 27B shows the records of only that portion of the experiment in which conditioning subthreshold stimulation of the TN was used. The behavior of the M-responses in this series was identical to that observed in the previous series. During simultaneous paired stimulation with conditioning subthreshold stimulation of the TN, the amplitude of the M-response of the Sol and ATM was significantly greater (trace 3) than during the single conditioning stimulation of the TN (trace 2) (the single test stimulus by itself did not evoke M-responses from the Sol and ATM--trace 1). During paired stimulation with a 1-msec delay, the M-responses of the Sol and ATM were smaller (trace 4) than during paired simultaneous stimulation, but were greater than for a single conditioning stimulation (trace 2). During paired stimulation with intervals of 2 or more msec (traces 5-11), the M-responses do not differ from those for a single conditioning stimulation.

As with the M-responses, the simultaneous subthreshold conditioning and test stimulation of the TN evokes an H-response from the Sol with a significantly greater amplitude (trace 3) than for a single test stimulus (trace 1). During paired stimulation with a 1-msec delay, this increase in the amplitude of the test H-response of the Sol over the control level is substantially reduced (trace 4). For interstimulus delays of 2 msec and more, the amplitude of the test H-response does not differ significantly from the control. During such delays between the PN conditioning stimulus and the TN testing stimulus, a sharp decrease is observed in the amplitude of the test H-response of the Sol (cf. the two curves in Fig. 27C).

Thus, the subthreshold stimulation of the TN evoked by the weak, direct stimulation of this nerve or by current influx during intense stimulation of the PN may actually be responsible for the increase in the H-response of the Sol observed during the simultaneous stimulation of the PN and the TN or when there is a 1-msec delay between stimuli. However, such a subthreshold stimulation of the TN does not produce a decrease in the H-response of the Sol if it precedes the test stimulation of this nerve by 2 or more msec.

The facts obtained experimentally make it possible to conclude that the short-latency decline in the test H-response of the Sol evoked by the conditioning stimulation of the PN is a result of reciprocal antagonist inhibition and is determined by spinal mechanisms.

It is not likely that antidromic effects via the stimulated PN motor fibers are a significant factor in the origin of the inhibitory effect. Investigations by Thomas and Wilson (1967) and Hultborn et al. (1971) have demonstrated that such antidromic volleys may exert a facilitatory, but not an inhibitory, influence on antagonist motoneurons. It remains unclear whether the effect described by us is analogous to Lloyd's "direct" reciprocal inhibition, which is known to be associated with the monosynaptic activation of the inhibitory interneurons via the I-a afferents of the antagonist (Eccles, J. C., 1957, 1969).

In testing the state of spinal reciprocal inhibition in man, Japanese investigators (Mizuno et al., 1971) have also employed the paired stimulation of antagonist nerves: conditioning stimulation of the PN and test stimulation of the TN. As their conditioning stimulus, they utilized rhythmic stimulation by 3 impulses with a frequency of 300 imp/sec and a strength up to 1.4 times the threshold for the M-response of the ATM. Mizuno and his colleagues found that in the healthy adult subject, such conditioning stimulation of the PN depresses the reflex excitability of the motoneuron pool of the GM in two phases--an early, weak depression and a late, stronger depression of the H-reflex of the GM. The latency of the early depression is approximately 7 msec, with a maximum after 20 msec (interval from the last of three conditioning impulses to moment of test stimulation); the late depression begins after 60 msec, with a maximum after 130 msec. During the early phase, the maximum decline in the test H-response of the GM is about 18% of the control response (magnitude of control response not indicated), and as much as 50% during the late phase. The degree of depression of the test reflex is, to a certain extent, directly dependent on the strength of the conditioning stimulus.

Since reciprocal inhibitory effects are also observed for a conditioning stimulus strength to the PN that is subthreshold for the M-response of the ATM, the Japanese investigators reason that the phenomenon investigated by them is analogous to Lloyd's I-a reciprocal inhibition. However, the minimum latency of the inhibitory effect in the study by Mizuno and co-workers is significantly greater than Lloyd's "direct inhibition" and permits the possible participation of other mechanisms triggered by the rhythmic activation of the I-a afferents of the antagonist.

In contrast to healthy subjects, in patients with bilateral athetosis, the Japanese investigators found a marked short-latency reciprocal inhibition with a latency on the order of 1-2 msec and with a maximum of between 1.5 and 3 msec. Mizuno and co-workers attribute the absence of short-latency inhibition in healthy subjects to the fact that the I-a reciprocal inhibitory pathway is apparently inhibited by supraspinal tonic influences under conditions of rest. In athetotic patients, there is a release from these tonic descending inhibitory influences.

With the conditioning stimulus strengths to the PN used in our experiments, the possibility cannot be disregarded that group II, as well as group I, afferents will be excited; and as experiments on animals have shown (Eccles, R. M., and Lundberg, 1959b; Holmquist and Lundberg, 1961), the group II afferents may evoke the reciprocal inhibition of extensor motoneurons. We thus have no basis for drawing a complete analogy between the phenomenon of the short-latency inhibition of the antagonist motoneuron pool in man and Lloyd's "direct" inhibition. However, as in "direct" inhibition, the reciprocal inhibition of antagonist motoneurons evoked by a volley in the group II afferents is associated with the activity of spinal inhibitory interneurons (Lundberg et al., 1962).

Although we are unable to completely identify the class of spinal interneurons responsible for the reciprocal inhibitory effect, we have reason to believe that the short-latency reciprocal inhibition of the GM motoneuron pool evoked by stimulation of the antagonist PN is associated with the activity of spinal inhibitory interneurons. We shall henceforth refer to this interneuronal spinal inhibitory apparatus as the SRIA. Under otherwise equal conditions, the degree of reciprocal inhibition of the antagonist is determined by the state of the SRIA. Hence, the state of the SRIA may be determined by the degree of depression of the test H-response of the Sol conditioned by preliminary stimulation of the antagonist PN.

State of the SRIA of the Direct Antagonist

To determine the state of the SRIA during the latent period and at the onset of voluntary foot flexion, we employed the paired-stimuli method of testing--conditioning stimulation of the PN and test stimulation of the TN, with a conditioning-testing interval of 3 msec.

As was indicated earlier (see Chapter 3, Section 2), if the voluntary EMG volley immediately follows the test H-response, it cannot be determined whether the reflex response of the motoneurons coincides with the onset of their voluntary impulse activity or precedes it by 10 msec or less. Neither is it possible to pinpoint the position of the test response during the first 30 msec of the voluntary EMG volley of the agonist ATM, since during paired-stimulus testing, a strong M-response of the ATM is evoked with a duration of 15-20 msec, followed by the "concomitant" H-response of the ATM (Fig. 28A). These responses "blend" with the voluntary EMG volley of the ATM (see traces 9 and 10). Thus, the position of the test H-response of the Sol can be determined with precision only if more than 30 msec have passed since the onset of the EMG volley of the ATM. In our experiments, calculations were made of the amplitudes of the Sol H-responses 40-60 msec after the onset of the voluntary EMG volley of the ATM. The average amplitude of the test H-response of the Sol 60 msec or more prior to the onset of the voluntary EMG volley of the ATM was assumed to be 100%, and the average amplitudes of the H-responses in other intervals were expressed as a percentage of this amplitude.

Portions of the recordings from one experiment are shown in Fig. 28A, and the average results of all 14 experiments involving 7 subjects are shown graphically in Fig. 28B, curve 2. For purposes of comparison, the average results from 46 experiments in which the latent period and the onset of the same movement (flexion of foot) were tested by a single H-response from the Sol are given in Fig. 28V, curve 1). All 7 subjects participated in both series of experiments (with single and paired stimuli).

As indicated by the data obtained from the experiments, during all but the last 30 msec of the latent period of the voluntary movement, the amplitude of the conditioned H-response of the antagonist Sol is constant. During the last 30 msec of the latent period, it is sharply reduced, and continues to decline after the onset of voluntary impulse activity by the motoneuron pool of the agonist ATM. The decline in the conditioned H-response of the Sol (antagonist) during the last 30 msec of the latent period cannot be explained by a depression of the reflex excitability of the motoneurons of this muscle, because--as shown in experiments with single-stimulus testing--such a depression does not occur

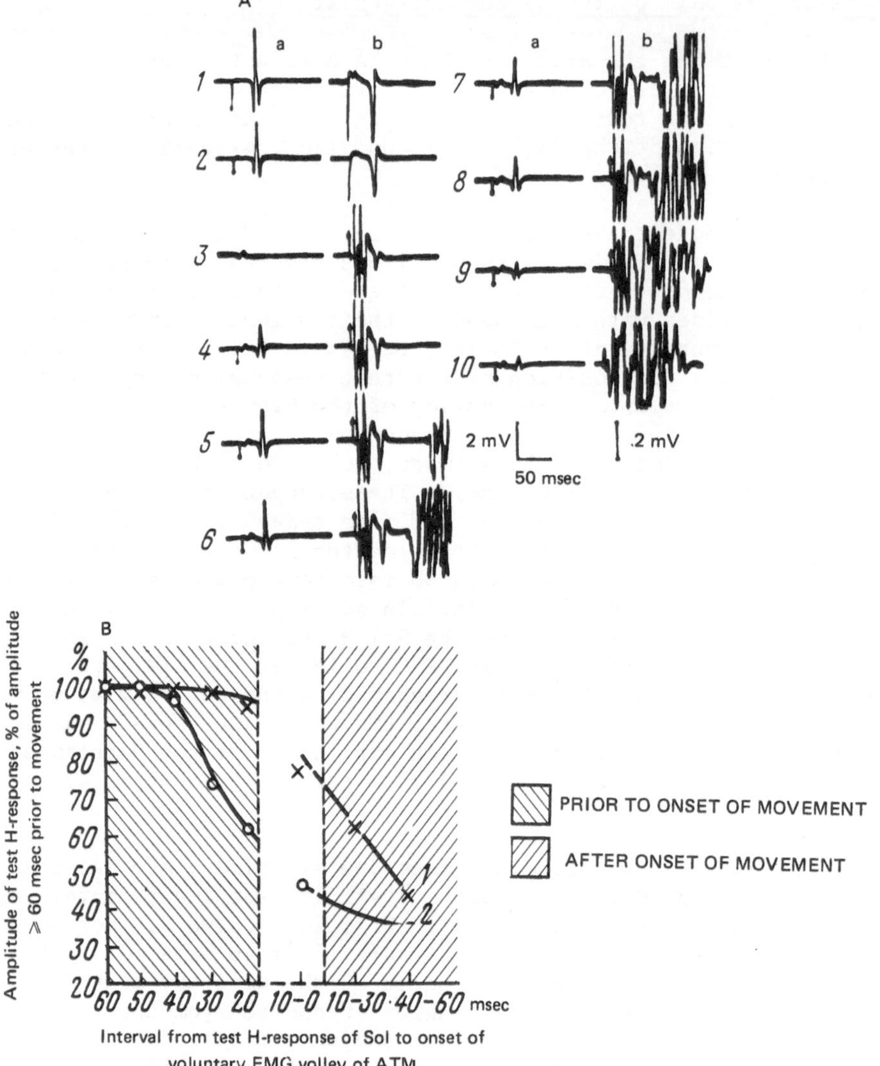

Fig. 28. Testing of latent period and onset of voluntary flexion of foot by paired stimulation of antagonist nerves. (A) EMGs of Sol (a) and ATM (b); (1-4) at rest: (1) maximal H-response of Sol; (2) single test stimulation of TN; (3) single conditioning stimulation of PN; (4) paired stimulation with 3-msec delay; (5-10) paired stimulation (3-msec delay) with test H-response of Sol prior to (5-8) and during (9 and 10) voluntary EMG volley of ATM. (B) Changes in amplitude of single (1) and conditioned (2) H-response of Sol during latent period and at onset of voluntary foot flexion.

during this period. It must therefore be concluded that the decline in the conditioned H-response of the Sol during the last 30 msec preceding the onset of voluntary movement is associated with an enhancement of the reciprocal inhibitory effect produced by the conditioning stimulation of the PN. Such an enhancement may be attributed to the facilitation of the SRIA. Since this facilitation occurs before the onset of voluntary movement, it is obviously determined by supraspinal influences.

Data obtained in a number of later studies are consistent with our results. Angel et al. (1970) discovered that in man, the duration of the silent period evoked by unloading of the voluntarily contracted pectoralis muscle is significantly lengthened if the onset of unloading is preceded by a light stimulus signaling the subject to relax the pectoralis muscle. This led Angel and his co-authors to conclude that changes occur in the excitability of spinal reflexes during the latent period of voluntary movement.

Tanaka (1972) also employed the paired-stimulus method described above (Mizuno et al., 1971) under conditions of the voluntary contraction of the foot flexors (constant pressure and direction of flexion movement). The maximal conditioned H-response of the GM does not change at rest, but diminishes during voluntary tensioning of the antagonists. Since inhibition is evoked by the subthreshold (for M-response) stimulation of the antagonist nerve and has a very brief latency (a few milliseconds), the author attributed this decline in the H-reflex of the antagonist to the effect of I-a reciprocal inhibition (see our criticism above). According to Tanaka, this mechanism is inhibited under conditions of rest, but during voluntary muscular contraction, the higher motor centers disinhibit it by activating the I-a reciprocal inhibitory interneurons. According to our findings, the supraspinal activation of the SRIA occurs prior to the onset of voluntary movement.

State of the SRIA of the Motoneuron Pool of an Inactive Muscle of an Adjacent Joint

To determine the extent to which the supraspinal "tuning" of the SRIA of the future antagonist of a voluntary movement is "specific," or essential for the organization of a concrete movement, we performed experiments designed to reveal the state of the SRIA of the GM motoneuron pool during the latent period and at the onset of voluntary movement in the hip joint, i.e., a movement in which the GM plays no active role. The conditions for the performance of this movement were such that no impulse activity of the tested GM motoneuron pool occurred during at least the first 60 msec of the voluntary impulse activity of the motoneuron pool of the primary agonist (hip flexor) (Zhukov, 1971).

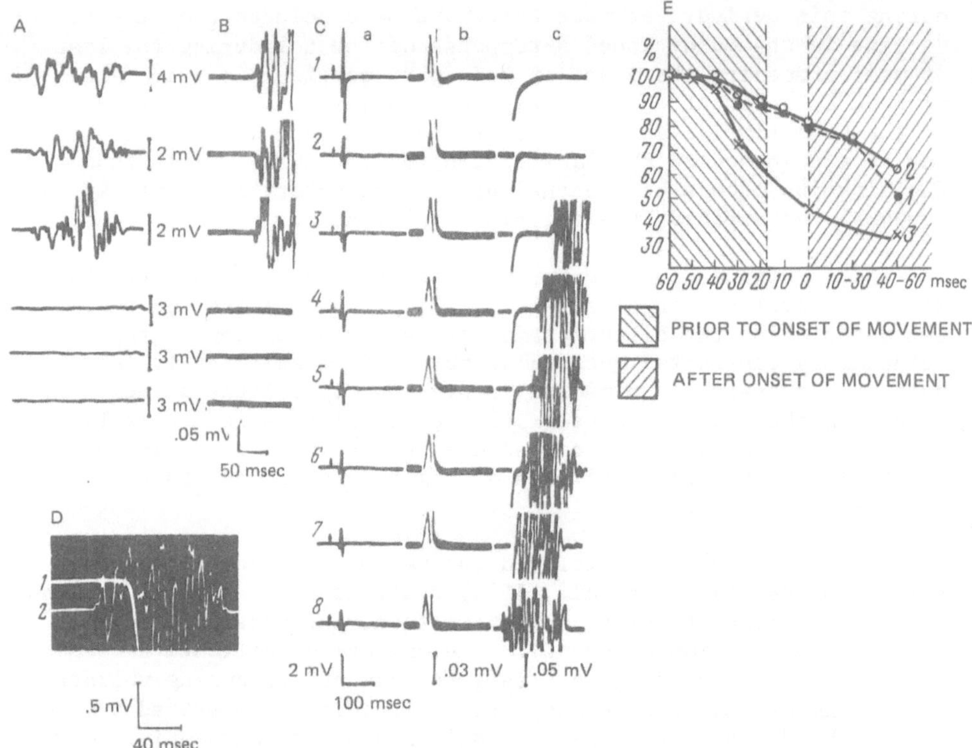

Fig. 29. EMG activity of lower limb muscles during voluntary flexion in hip joint (A, B, D) and H-response of Sol during latent period and at onset of this movement (C). (A,B) From top to bottom: m. sartorius; m. tensor fasciae latae; m. rectus femoris; m. semitendinosus, Sol, and ATM. (C) EMGs of Sol (a), ATM (b), Sart (c); (1-2) maximal and control H-responses of Sol at rest; (3-8) H-response prior to (3-5), accompanying (6), and after onset of (7,8) voluntary EMG volley of hip flexors. (D) Goniogram of hip joint (1) and EMG of Sart (2). (E) Changes in amplitude of single (1) and conditioned H-response of Sol (2) during latent period and at onset of voluntary flexion in hip joint; (3) conditioned H-response of Sol during latent period and at onset of voluntary flexion of foot.

The EMGs of the muscles of the ankle, hip, and knee joints during flexion in the hip joint are shown in Figs. 29A, B, and D. In Figs. 29A and B, the three upper records show the EMG activity of the primary agonists of the movement; the two lower records, that of the antagonist muscles of the ankle joint--the Sol and ATM. The simultaneous recording of the electrical activity of the primary agonist of the given movement and the goniogram of movement in the hip joint are shown in Fig. 29D.

As the records in Fig. 29 indicate, the onset of the voluntary EMG volley of all the agonists of hip flexion coincide (see the three upper records under B). The average duration of the EMG volley of these muscles is approximately 150 msec (A). The electromechanical delay from the onset of the EMG volley of the primary agonists to the onset of flexion in the hip joint amounts to 40-60 msec (D). At least the first 60 msec of voluntary agonist activity is accompanied by no electrical activity by the Sol or ATM (B and C).

Figure 29C shows portions of the records from experiments involving the single-stimulus testing of the GM motoneuron pool during the latent period and at the onset of voluntary flexion in the hip joint. The average results of 11 experiments are shown graphically in Fig. 29E, curve 1: during the last 30 msec of the latent period preceding voluntary hip flexion, there is a decline in the reflex excitability of the motoneuron pool of the GM. This decline continues after the onset of the voluntary activity of the hip flexors, and grows more abrupt at more than 30 msec after the onset of this voluntary activity.

Thus, the voluntary impulse activity of the motoneuron pool of the hip flexors is preceded by a depression of the reflex excitability of the motoneuron pool of the extensor of the adjacent hip joint. Interestingly enough, these changes in the reflex excitability of the motoneuron pool of the inactive GM are observed during the last 30 msec of the latent period of the voluntary movement, i.e., in the interval characterized by abrupt, supraspinally evoked ("triggering") changes in the reflex excitability of the "fast" motoneurons of the primary agonist (see Chapter 3, Section 2), in the state of the interneuronal inhibitory apparatus acting on the primary agonist (see Chapter 3, Section 4), and in the state of the interneuronal reciprocal inhibitory apparatus acting on the primary antagonist (see above).

Thus, it seems quite probable that changes in the reflex excitability of the motoneuron pool of the muscle of the adjacent joint that are observed during the latent period of the voluntary movement are determined by the same supraspinal command that determines the other changes in the segmental neuronal motor control apparatus described above.

The supraspinal command for a voluntary movement apparently "triggers" the same spinal system of long interrelations between the motoneuron pools of adjacent muscles that can also be triggered by reflex stimulation (Creed et al., 1932). This assumption would account quite well for the similarity in the distribution of excitatory and inhibitory influences to the motoneuron pools of adjacent joints according to their functional similarity that is observed in the phenomenology of long reflexes in spinal preparations (Creed et al., 1932) and that, by our findings, is manifested in a change in the reflex excitability of motoneuron pools of muscles of adjacent joints during the organization of voluntary movement (Kots and Zhukov, in preparation).

To determine the state of the SRIA acting on the motoneuron pool of an inactive muscle of an adjacent joint during the latent period and at the onset of voluntary flexion in the hip joint, we tested the reflex excitability of the Sol motoneurons by the paired-stimulus method. The average results of our experiments are shown graphically in Fig. 29E, curve 2. As the data indicate, the amplitude of the conditioned H-response of the Sol does not change prior to the last 30 msec of the latent period of voluntary flexion in the hip joint. During these last 30 msec and the first 30 msec of the voluntary EMG volley of the hip flexors, the amplitude of the conditioned H-response is reduced below the control value; this reduction intensifies after the first 30 msec of the voluntary EMG volley. It is most significant that the relative decline of the conditioned H-response of the Sol corresponds exactly to the relative decline of the single, nonconditioned H-response of the Sol (curve 1). This means that in the first case, the decline in the Sol H-response can be determined entirely by the decrease in the reflex excitability of the motoneurons, but not by a change in the state of the SRIA.

Hence, in contrast to the supraspinal "tuning" of the SRIA acting on the motoneuron pool of the future direct antagonist, during the organization of voluntary movement, there is no supraspinal facilitation of the SRIA acting on the motoneuron pool of the inactive muscle of the adjacent joint. Inhibition of the reflex excitability of the latter prior to and during the onset of voluntary movement is of a different nature, and is not associated with the increased activity of its SRIA.

Thus, on the basis of our findings, it can be concluded that the organization of a phasic voluntary movement is accompanied by the specific supraspinal facilitation ("tuning") of the SRIA of the future direct antagonist of the movement in the given joint. It is quite likely that during the performance of other movements of the ankle joint or in other variants of foot flexion, there would be changes in the degree and character of supraspinal "tuning" influences on the same SRIA, and possibly changes in the sign (direction) of the influence--from facilitation of the SRIA, as in the present case, to inhibition, as when the simultaneous activity of a muscle pair is required in the performance of a movement or postural fixation.

The supraspinal regulation of the state of the SRIA is apparently one of the mechanisms by wich the higher motor centers determine the interaction of the spinal centers of antagonist muscles that will provide for the performance of a required movement.

Pursuant to such a hypothesis, it was important to determine whether the character of the supraspinal "tuning" of the SRIA changes during the refinement of voluntary motor performance as a result of systematic training.

Changes in the Supraspinal "Tuning" of the SRIA as a Result of Training

In a large number of studies, the electromyographic method has been used to investigate changes in the level and ratio of the activities of antagonist muscles occurring as a result of the systematic repetition of a learned movement. According to one investigator, changes in the ratio of the electrical activities of antagonist muscles occur by a reduction in the electrical activity of the antagonist; according to another investigator, by diametric changes in the activity of the antagonist muscles--an increase in agonist activity and a decrease in antagonist activity (Person, 1965). Such contradictions in the findings of different investigators may be attributable to differences in the joints of the subjects tested, to variations in the complexity of the learned movements, or to the choice of different pairs of antagonist muscles.

However, there is one methodological error in studies such as these that has been often overlooked in the past, namely, that the electrical activity levels of a muscle of the same subject over a period of several days can be compared only by the careful reproduction of recording and amplification conditions from one day to the next, a precaution that is not taken in all cases. Even in cases in which these conditions are adequately monitored, however, it is apparently very difficult (if not impossible) to create standard conditions under which the electrical activity of a given muscle can be compared on different days of the training period.

Mindful of these pitfalls, we performed an electromyographic study on the dynamics of the electrical activity of the antagonist muscles of the ankle joint and their activity ratio during the systematic repetition of a learned movement (flexion of foot against weight), trying if possible to avoid the methodological shortcomings mentioned above. In addition, during the training cycle, we studied changes in the dynamics of the reflex excitability of the motoneuron pool of the Sol (the future antagonist) and in the character of the supraspinal "tuning" of the SRIA during the organization and at the onset of this voluntary movement.

The training consisted of the daily, multiple repetition of a rapid foot flexion, from its initial resting position (90° in the ankle joint), against a 2-kg weight. The movement was repeated 200-250 times on each training day. The subject was given a brief rest period between training periods to avoid fatigue. There were 9 training days in all. The electrical activity of the antagonist muscles was recorded during the performance of the first 10 movements from 3-6 times during the training cycle: once on the first day and once on the last day, with several recordings in between. After recording, the latent period and the onset of the trained movement were tested by the monosynaptic testing method (single- or paired-stimulus).

To standardize as much as possible the conditions under which the electrical activity of the antagonist muscles was recorded, a constant interelectrode distance was maintained, the electrode positions were marked, an attempt was made to keep the interelectrode resistance constant (by treating the skin), and a "Diza" electromyograph with a high input resistance was used. But our principal method of ensuring standard recording conditions was to determine not the absolute (voltage), but the relative, indices of muscular activity. To do this, at the beginning of the experiment, the subject was given 2 sec in which to fix the angle of the ankle joint at 45° while supporting a 2-kg weight on the tested foot. Then the electrical activity of the Sol and ATM was recorded over a period of 0.5 sec. This electrical activity was taken as our "standard." The electrical activity of each of the antagonist muscles during the performance of the trained phasic movement was determined as a fraction of the "standard" activity of this muscle, this being done as follows: The amplitudes of potential oscillations in the EMG of a given muscle were summed for a 300-msec period, and division by 5 yielded the average level of the standard activity of a given muscle for a 60-msec interval--the "standard activity measure" of the muscle. The electrical activity of the muscle during the performance of the trained phasic movement was calculated on the basis of the EMG of the first 10 repetitions of this movement. For each of these 10 EMGs, we summed the amplitudes of potential oscillations for the first 60 msec of the voluntary EMG volley and then computed the average of 10 such trials. The average electrical activity of a muscle (for the first 60 msec) during the performance of a phasic movement determined in this way was related to the "standard activity measure" of the muscle in question. We thus obtained a relative measure of the activity of a given muscle during the performance of a trained movement on different days of the training cycle. Besides a determination of the relative activity of the agonist (ATM) and antagonist (Sol), we also calculated their activity ratio--ATM/Sol.

The average indices of the relative activity of the antagonist muscles and their activity ratios during the performance of the trained movement by 6 subjects on different days are shown graphic-

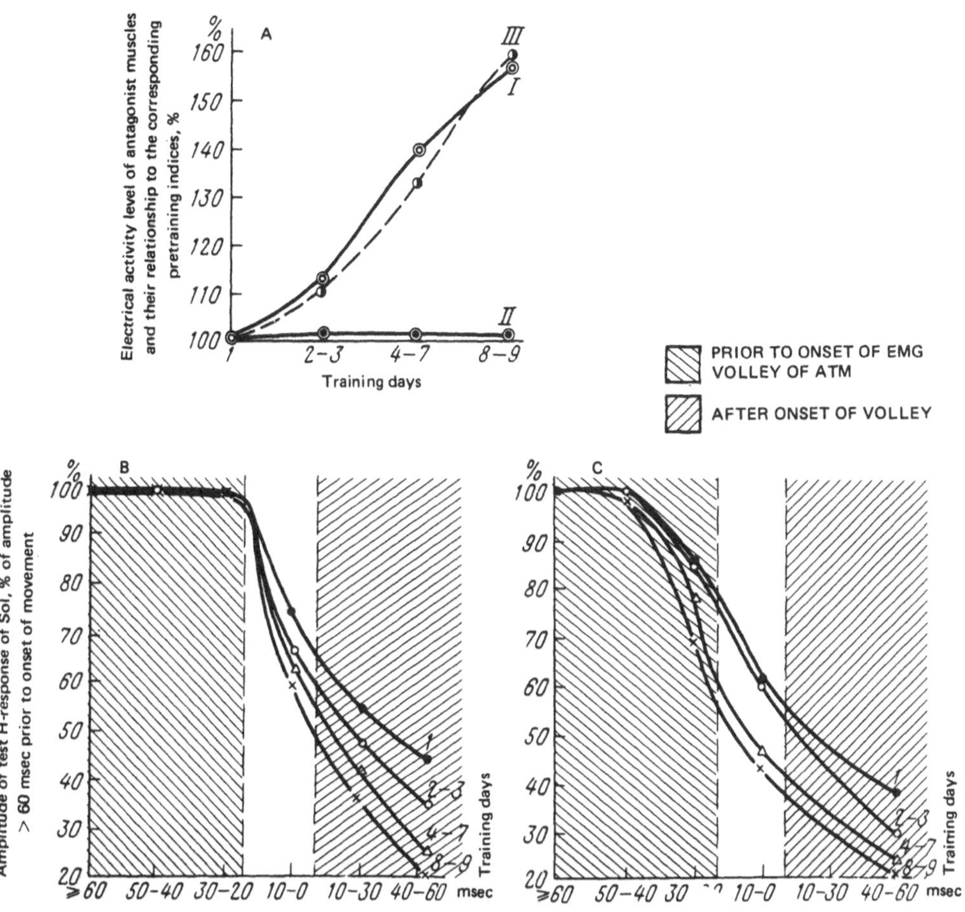

Fig. 30. Effect of training on the electrical activity level of antagonist muscles (A), the dynamics of the reflex excitability of the antagonist motoneuron pool (B), and the state of the SRIA (C) during the organization and at the onset of voluntary flexion of the foot. (A) Electrical activity of ATM (I) and Sol (II) and ATM/Sol activity ratio (III).

ally in Fig. 30A. In the course of the training cycle, the relative electrical activity of the agonist (ATM) increases significantly, and is, on the average, 1.5 times greater by the end of the training cycle than at the beginning. At the same time, the relative electrical activity of the primary antagonist (Sol) exhibits no changes during the course of the training cycle. Accordingly, the activity ratio of the antagonist muscles increases as a result of training such that by the 8th or 9th training day, the ATM/Sol ratio is, on the average, 1.6 times greater than prior to training.

Training also modifies the character of the electrical activity of the agonist ATM during different phases of the movement. If the EMG volley of the ATM is characterized by a gradual increase in oscillatory amplitudes toward the end of the movement during the first days of training, the character of the EMG volley is radically changed by the end of the training cycle: now even the onset of the volley is characterized by high-amplitude oscillations, which gradually diminish toward the end of the movement.

As the experiments with H-reflex single testing have shown, training has no effect on the dynamics of the reflex excitability of the motoneuron pool of the future antagonist during the latent period of voluntary movement. The average results from experiments of this series are shown graphically in Fig. 30B. It can be seen that both before and after training, the reflex excitability of the motoneuron pool of the future antagonist does not change prior to the onset of voluntary activity.

Training leads to a significant increase in the reciprocal inhibition of the antagonist motoneuron pool during the voluntary impulse activity of the agonist motoneuron pool. For example, the depression of the test H-response of the antagonist Sol in the intervals 10-0, 10-30, and 40-60 msec after the onset of the voluntary agonist EMG volley is already increased by the 2nd or 3rd day of training. This increase becomes significant by the 4th to 7th day of training, and continues to grow through the 8th and 9th (final) days. The greater the interval from the onset of the voluntary EMG volley to the moment at which the H-response of the antagonist is evoked, the more pronounced is the effect of training in enhancing reciprocal antagonist inhibition. Thus, on the 8th to 9th day of training, the average amplitude of the H-response of the Sol in the 10- to 0-msec interval is about 80% of the H-response of the Sol in the same interval at the start of training; in the 10- to 30-msec interval after the onset of the EMG volley, this ratio falls to 65%; and in the 40- to 60-msec interval, to 45%.

Thus, experiments in which we recorded the electrical activity of antagonist muscles and tested the reflex excitability of the antagonist motoneuron pool on different days of the training cycle have demonstrated that training leads not only to the enhancement of the facilitatory influences on the agonist motoneuron pool (increase in EMG of ATM), but also to an increase in reciprocal inhibitory influences on the antagonist motoneuron pool that was observed to accompany the onset of voluntary impulse activity of the agonist, but not during the latent period preceding this activity.

In light of these results, it appeared useful to trace the dynamics of changes in the state of the SRIA as training progressed. To do this, on different days of the training cycle, we tested the latent period and the onset of the trained voluntary movement by the method of paired stimulation of the antagonist nerves (see above). The average results of the experiments (7 subjects) are shown graphically in Fig. 30C.

As during unresisted flexion of the foot ("standard movement"), the conditioned H-response of the Sol does not change during all but the last 30 msec of the latent period preceding the onset of flexion of the weighted foot. There is a considerable depression of this H-response during this interval, which intensifies at the onset of voluntary agonist activity. The training process is accompanied by a gradual enhancement of this phenomenon, such that by the 8th or 9th training day, the amplitude of the conditioned H-response of the Sol is significantly smaller during the last 30 msec of the latent period and the first 40-60 msec of voluntary activity than in the same intervals at the start of the training cycle.

The training-related increase in the inhibitory effect as revealed by paired-stimulus testing is relatively less pronounced in the 30- to 20-msec interval prior to movement, and grows increasingly more prominent as the onset of voluntary movement draws near, as well as during the onset of movement. For example, the amplitude of the conditioned H-response of the Sol by the 8th or 9th day of training is about 80% in the interval 30-20 msec prior to movement, about 70% in the interval 10-0 msec, and about 50% in the interval 30-40 msec after the onset of movement, relative to the corresponding indices at the start of training.

As was mentioned earlier, training produces no noticeable changes in the reflex excitability of the future antagonist motoneuron pool at any time during the latent period: both prior to training and after 8 or 9 days of systematic training, the reflex excitability of the antagonist motoneuron pool remains roughly the same. At the same time, paired-stimulus testing reveals inhibitory effects as early as during the last 30 msec of the latent period, both before and after training (though these effects are considerably enhanced after training). The results obtained lead us to conclude that one result of training is an increase in the supraspinal facilitation ("tuning") of the SRIA of the future antagonist of a voluntary movement. This in turn leads to an increase in the reciprocal inhibitory influences acting on the antagonist motoneuron pool, which arises during the period of the voluntary impulse activity of the agonist motoneuron pool.

From our point of view, the facts obtained in the present experiments indicate that one of the mechanisms by which the voluntary control of antagonist muscles is refined is a gradual refining of the process of the organization of movement, which includes the formulation of an increasingly adequate supraspinal command. It is significant that the "address" of this command at the spinal level appears to be the interneuronal segmental apparatus by which the interaction of the motoneuron pool of antagonist muscles is probably realized during a voluntary movement.

CHAPTER 5

DESCENDING REFLEX INFLUENCES DURING THE ORGANIZATION OF VOLUNTARY MOVEMENT

In the previous chapters, we have described the results of experiments that revealed complex changes in the spinal motor apparatus generated by supraspinal influences during the organization of voluntary movement. It is important to ascertain which structures of the brain determine these influences, and by which descending pathways they are mediated.

One method of identifying the neural structures involved in the organization of voluntary movement is to study the changes in the segmental motor apparatus evoked by specific supraspinal reflex influences. In this regard, it proved methodologically convenient to investigate vestibulospinal influences (Section 1). We also investigated the spinal changes evoked by descending interlimb reflex influences (Section 2). Although the structures that determine interlimb reflex influences are still unknown, our investigations have nevertheless shown that different descending pathways are undoubtedly used for this type of influence than for the vestibulospinal influences.

Thus, we studied the changes in spinal motor centers evoked by two different types of supraspinal reflex influences mediated via different descending pathways. A determination of the similarities and differences in spinal effects evoked by reflex descending and voluntary influences was used in analyzing the pathways and mechanisms by which a voluntary supraspinal command is realized.

Another technique for resolving the last problem is to study the interaction of the influences of different descending motor systems during the organization of voluntary movement. To this end, we studied the spinal effects of the interaction of voluntary, vestibulospinal, and descending interlimb reflex influences during the latent period and at the onset of voluntary movement (Section 3).

1. VESTIBULOSPINAL REFLEX INFLUENCES DURING REST

Our knowledge of the structural and functional organization and the mechanisms of the vestibulospinal system has been derived almost entirely from experimentation with cats. The vestibulospinal projections in the cat may be divided into two systems: (1) the vestibulospinal tract and (2) vestibular fibers descending in the medial longitudinal fasciculus and its continuation in the spinal cord (Nyberg-Hansen, 1964). In contrast to the predominantly dorsolateral position of the corticospinal and corticorubrospinal fiber endings in the intermediate area of the spinal cord (in lamina IV and V-VII), both vestibulospinal fiber systems (as well as the reticulospinal tract) terminate more ventromedially (in lamina VI-VIII and IX), among the interneurons that project predominantly to the motoneurons of axial and proximal limb musculature (Nyberg-Hansen, 1964, 1966; Rexed, 1964; Romanes, 1964). Besides their common endings in the spinal cord, the vestibular and reticular systems also have close connections at the level of the brain stem (Brodal et al., 1962; Ladpli and Brodal, 1968). It is therefore quite likely that the reticulospinal tracts supplement the vestibulospinal tracts in the transmission of impulses of vestibular origin to the spinal cord (Gernandt and Thulin, 1955). Vestibular influences mediated by the reticular formation are apparently transmitted over both the crossed and uncrossed reticulospinal tracts (Brodal et al., 1962). The close association and functional similarity of the vestibulo- and reticulospinal descending systems have led Kuypers to regard them as a single medial system (Kuypers, 1964).

On the basis of its functional characteristics, the vestibulospinal system comprising Kuyper's medial system is now considered to be the functional antagonist of the corticospinal ("lateral," after Kuypers) system (Kostyuk, 1973). The vestibulospinal system essentially controls the axial and proximal limb musculature (particularly the extensors) and is involved in the regulation of posture and locomotion, while the corticospinal (pyramidal) system and the corticorubrospinal system (which comprise Kuyper's lateral system) are of special importance in the control of distal limb musculature (particularly the flexors) and govern discrete coordinated movements (Kuypers, 1964; Lawrence and Kuypers, 1968a,b; Kostyuk, 1970, 1973).

The consequences of the acute or chronic destruction of the vestibular nuclei (Sprague and Chambers, 1953), transection of the VIIIth nerve (Batini et al., 1957), transection of the vestibulospinal tract in animals (Gernandt and Thulin, 1953) and in man (Hyndman, 1943), extirpation of the labyrinth in man (Lorento-de-No, 1931), and electrical stimulation of the vestibular nuclei (Sprague et al., 1948; Pompeiano, 1960) have shown that the vestibulospinal system has predominantly ipsilateral connections that facilitate extensor motoneuron pools.

There is, however, a complex bilateral interaction of vestibulospinal influences. For example, as early as 1927, Spiegel (1927) observed that following unilateral destruction of the vestibular nuclei, the supplementary destruction of the contralateral nuclei is accompanied by the restoration of decerebrate rigidity on the side of the initial destruction. Similar effects have been observed following the successive bilateral transection of the VIIIth nerve (Batini et al., 1957). Judging by the results of the experiments by these investigators, the reduction of ipsilateral rigidity subsequent to unilateral transection of the VIIIth nerve is associated not only with the termination of facilitatory labyrinthine influences, but also with the inhibitory action of afferentation from the vestibular receptors of the opposite side. Transection of the contralateral VIIIth nerve is followed by hypertonus of previously flaccid forelimb extensors.

Very little is known about the organization of the vestibulospinal system in man. There are data indicating that the vestibulospinal tract in man (and in the chimpanzee) is relatively poorly developed in comparison with that in the cat and the lower monkeys (Schoen, 1964). At the same time, the number of nerve and glial cells in all the vestibular nuclei in man is significantly greater than in the dog and lower monkeys (Blinkov and Ponomarev, 1965). Sadjadpour and Brodal (1968) investigated the cytoarchitectonics and fiber structure of the vestibular complex in four human subjects and found a similar organization in the vestibular nuclei of man and of the cat.

It is quite probable that in man, as in other mammals, vestibulospinal influences are mediated by an independent descending system with localization and mechanisms of spinal influences quite distinct from those of the corticospinal system. However, the mechanisms of these influences have not yet been studied in humans.

To study the vestibulospinal mechanisms, we considered it useful to employ the method of electrical stimulation of the vestibular apparatus; this method makes it possible to accurately control the intensity and duration of the stimulus, accurately compute the time following the stimulus, and selectively excite one of the labyrinths (for a survey, see Kots, 1972). Other methods of stimulating the vestibular apparatus in man (angular acceleration, caloric method) do not have all these advantages.

The effect of vestibular stimulation was determined by the change in the amplitude of the control H-response of the GM evoked after a predetermined delay following the onset of electrical vestibular stimulation. The amplitude of the control H-response (H_c) was about 40% of the maximal response. A special index was used to quantitatively evaluate the vestibulospinal effect--the "vestibular addition" (VA), i.e., the percentage ratio of the difference between the amplitude of the H-response evoked after vestibular stimulation (the "vestibular" H-response, or H_v) and the amplitude of the control ("pure") H-response (H_c) to the amplitude of the control H-response:

$$VA = \frac{H_v - H_c}{H_c} \cdot 100\%$$

During the experiment, the subject was seated in an armchair. The test H-response was evoked at approximately 10-sec intervals. After 2-6 control responses, a "vestibular" H-response was evoked. Thus, the vestibular stimulus was applied 30-80 sec after the preceding stimulus at a moment unknown to the test subject. The experiments showed that the amplitude of a control H-response evoked 10 sec after the "vestibular" H-response is usually greater than at rest with no vestibular stimulation (vestibular aftereffects), while subsequent H-responses--i.e., approximately 20 and more msec after the vestibular stimulation--do not differ from the H-response under control conditions. Taking this circumstance into account, we used the difference between the amplitude of the "vestibular" H-response and the amplitude of the last control H-response generated immediately prior to this "vestibular" H-response in determining the VA.

Since the vestibular stimulation was repeated up to several dozen times during the course of any one experiment, we compared the VAs obtained during the first and second halves of the experiment. No adaptation was observed with respect to vestibulospinal effects: when a constant vestibular stimulus strength is used, the average VA in the second half of the experiment does not differ from the VA in the first half. This made it possible to determine average data by the results of the entire experiment.

During stimulation of the ipsilateral vestibular apparatus by a 1-msec anodal (Fig. 31A) or cathodal current, a statistically significant increase in the test H-response begins at a 30-msec delay between the onset of the vestibular and test stimulations. As the delay is increased, the amplitude of the test H-response increases smoothly for about 60 msec, i.e., up to a 90-msec interstimulus delay. At such a delay, the vestibular" H-response achieves a maximum and is approximately twice as strong as the control response; i.e., the VA is approximately 100%. When the delay is increased above 90 msec, the VA declines rapidly, such that by a 140-msec delay, the VA is only about 14%. The slight differences between anodal and cathodal effects are probably due to the use of a nonuniform current strength (Kots and Mart'yanov, 1967).

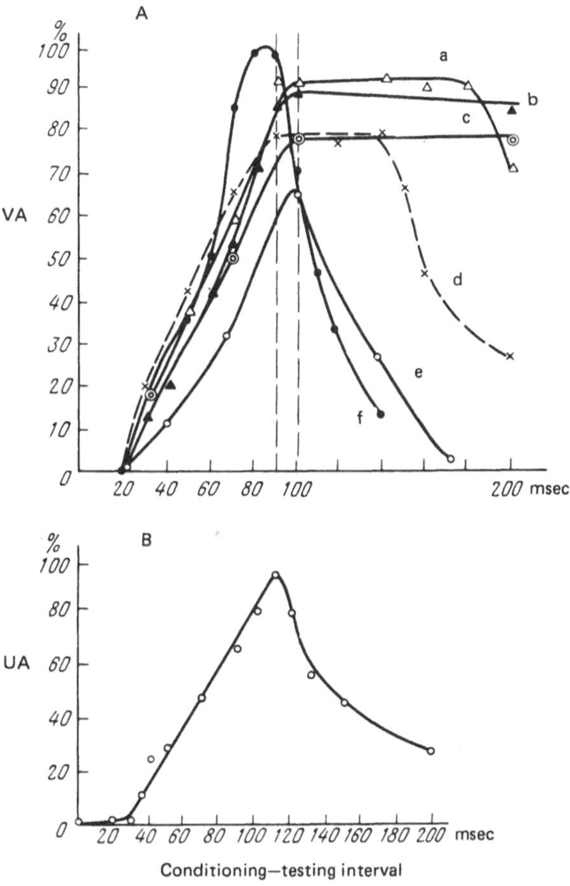

Fig. 31. Time course of ipsilateral vestibulospinal (A) and descending interlimb (B) reflex facilitatory effects. (a) Anode, 30 msec; (b) anode, constant; (c) cathode, constant; (d) anode, 10 msec; (e) cathode, 1 msec; (f) anode, 1 msec.

Increasing the duration of the vestibular stimulation has almost no effect on the time course of the vestibulospinal facilitatory effect for the first 90-100 msec after the onset of vestibular stimulation (see Fig. 31A). During the 10- and 30-msec anodal or cathodal stimulation of the vestibular apparatus, as well as the galvanic cathodal or anodal stimulation (duration more than 0.1 sec), the first significant increase in the test H-response is observed after a 30-msec delay, and during the next 60-70 msec, the amplitude of the "vestibular" H-response increases continuously with increasing delay. The maximum VA is recorded at delays of 90-100 msec after the onset of vestibular stimulation, and may reach 100% for a high stimulus strength to the vestibular apparatus.

Unlike the effect of 1-msec stimulation following 10-msec (and longer) vestibular stimulation, the maximum vestibulospinal facilitatory effect disappears quite slowly. The length of the plateau of the maximum facilitatory effect is proportional to the duration of vestibular stimulation (see Fig. 31A). Following cathodal stimulation of duration greater than 1 sec (galvanic stimulation), the maximum VA is maintained for 0.3 sec after the onset of vestibular stimulation, and even 1 sec after the onset of vestibular stimulation, the VA is still about 70% of the maximum VA. After galvanic anodal stimulation, the plateau is somewhat shorter than after cathodal stimulation.

Thus, regardless of the duration and polarity of the vestibular stimulation, the time course of the vestibulospinal facilitatory effect is the same during the first 90-100 msec after the onset of vestibular stimulation: beginning about 30 msec after the onset of vestibular stimulation, there is a steady increase in the reflex excitability of the ipsilateral (relative to stimulated vestibular apparatus) motoneurons of the GM, lasting 60-70 msec. After attaining a maximum, the reflex excitability of the motoneurons either begins an immediate decline or exhibits a plateau value for some time before beginning a gradual decline. The greater the duration of the electrical vestibular stimulation, the greater the overall duration of the facilitation period of the GM spinal motoneurons. The strength of vestibular stimulation has, within certain limits, a direct effect on the magnitude of the facilitatory effect and its overall duration, but not on the duration of the first phase, i.e., the phase characterized by a smooth increase in the reflex excitability of the GM motoneurons (Kots, 1972).

If a control H-response with an amplitude less than 50% of the maximal is used, the absolute increase in the "vestibular" H-response over the control response is inversely proportional to the magnitude of the control H-response, while the relative increase (i.e., the VA) is independent of the magnitude of the control H-response and amounts to approximately 100% for high stimulus strengths to the vestibular apparatus. If the amplitude of the control H-response is more than 50% of the maximal, the absolute increase in the "vestibular" H-response, like the VA, is inversely proportional to the proximity of the control response to the maximal value. The smallest "vestibular" changes in the H-response are observed if maximal and supramaximal H-responses are used as the control responses. The maximal "vestibular" H-response never exceeds the maximal control H-response by more than 10%.

The time course of the contralateral vestibulospinal facilitatory effect coincides essentially with that of the ipsilateral effect. In the first case, the first signs of an increase in the reflex excitability of the motoneuron pool of the contralateral GM are observed on evocation of a test H-reflex 30 msec after the onset of vestibular stimulation; the reflex excitability of the contralateral motoneuron pool then increases smoothly for the next 60-70 msec, such that after a 90- to 100-msec delay, the contralateral "vestibular" H-response achieves its maximal amplitude. The presence or duration of a plateau at this point and the time required for recovery of the initial amplitude are directly dependent on the duration of the vestibular stimulation.

The threshold for the contralateral vestibulospinal facilitatory effect is higher than for the ipsilateral effect. In each of the experiments (with a 100-msec delay from the onset of vestibular stimulation to test stimulation), we were able to choose a stimulus strength to the vestibular apparatus that produced a significant increase in the amplitude of the ipsilateral "vestibular" H-response over the control level, but that was inadequate to evoke a significant increase in the test H-response in the contralateral limb. An increase in stimulus strength above "subthreshold" leads to the emergence of a contralateral effect and to an increase in the ipsilateral effect. Further strengthening of the vestibular stimulus increases both the ipsi- and the contralateral VAs, but at relatively low stimulus strengths, a significant difference is maintained between the ipsi- and contralateral VAs. During strong cathodal (average 2.2 mA) or anodal (average 3.1 mA) stimulation of the vestibular apparatus, there are no noticeable differences between the ipsi- and contralateral VAs.

Thus, unilateral electrical vestibular stimulation induces the facilitation of the spinal motoneurons of the GM of both lower limbs. The ipsi- and contralateral facilitatory effects differ only in threshold and magnitude: the ipsilateral effect has a lower threshold and, at relatively low strengths of vestibular stimulation, a greater magnitude than the contralateral effect.

The results obtained are in agreement with the findings of a great many experimental (Sprague et al., 1948; Gernandt and Thulin, 1953; Pompeiano, 1960; Orlov, 1962; Brodal et al., 1962) and clinical (Lorento-de-No, 1931; Hyndman, 1943; Kalinovskaya and Yusevich, 1963) investigations, which revealed predominantly ipsilateral connections between the vestibular nuclei and the spinal motoneurons.

To account for the contralateral facilitatory effect, one must take note of the presence of crossed vestibulospinal pathways at the level of the spinal cord (Brodal et al., 1962), and "horizontal" connections between the vestibular nuclei and the homo- and contralateral nuclei of the reticular formation of the brain stem (Szenthagothai, 1943; Gernandt and Thulin, 1952; Ladpli and Brodal, 1968). These neural pathways ensure the transmission of vestibulospinal influences over the crossed and uncrossed reticulospinal pathways, as well as bilateral connections of the vestibular nuclei of the right and left sides (Bekhterev, 1905; Hogyes, 1912; Brodal et al., 1962; Gorgiladze, 1966; Leshchinyuk, 1968).

To determine the role of electrodermal stimulation in the origin of the effects described, we determined the changes in the amplitude of the test H-response 100 msec after the onset of electrical stimulation of the ipsilateral earlobe in 8 subjects (Mart'yanov, 1968). Higher stimulus strengths were used in stimulating the earlobe than in vestibular stimulation. The duration of the stimulating impulse was 10-30 msec or more than 0.2 sec. In each experiment, the average amplitude of the control H-response and the amplitude of the H-response following stimulation of the earlobe were determined on the basis of 20 responses. In all 8 experiments, the average amplitude of the conditioned H-response did not differ from the average amplitude of the control H-response. We concluded, therefore, that the facilitatory effects described are not related to electrodermal stimulation or to the voluntary reactions of the subjects to painful electrodermal stimulation.

Clinical physiological studies* were done to determine the site of action of galvanic vestibular stimulation. In three patients with a surgically transected left VIIIth nerve, the test H-reflex of the GM was evoked 100 msec after the onset of galvanic vestibular stimulation with stimulating electrodes attached to the ear and contralateral cheek, and ear and chest. With the active electrode attached to the left ear, no significant increase was observed in the test H-response of the ipsi- or contralateral GM in any of these patients. Only in one patient, with the stimulating electrodes attached to the ear and contralateral cheek, was a slight increase (statistically insignificant) observed in the test H-response of the contralateral (right) GM. Since no such increase was detected with the stimulating electrodes attached to the left ear and chest, it is likely that in the first case (ear-cheek), the increase in the H-response was associated with the stimulation of the healthy right labyrinth by the electrode on the cheek near the right ear.

In all three cases, stimulation of the healthy right ear was accompanied by a significant increase in the test H-response of both the ipsi- and contralateral GM. The contralateral vestibulospinal facilitatory effect accompanying stimulation of the intact

*The studies were done at the N. N. Burdenko Institute of Neurosurgical Research with the support of Dr. Med. Sci. A. R. Shakhnovich, and at the Department of Neurology of the S. P. Botkin State Clinical Hospital with the assistance of V. S. Mal'tsina.

vestibular apparatus in these patients was, on the whole, no weaker than the ipsilateral effect (Kots, 1972), a circumstance that is not usually observed.

On the basis of our results, it may be concluded that the reflex pathways of the facilitatory spinal effect evoked by galvanic stimulation of the ear include the VIIIth nerve (most likely its vestibular portion). In addition, these data tend to refute the hypothesis regarding the voluntary character of vestibulospinal facilitation, since the patients' subjective reactions to electrical stimulation (threshold and maximum endurable stimulation) were more or less the same, regardless of the ear stimulated. Moreover, in patients with "pyramidal" paralysis, distinct vestibulospinal facilitatory effects are observed during testing of the affected side even in the absence of voluntary movement in the ankle joint and voluntary "tuning" of the reflex excitability of the motoneurons of the GM on the affected side (see Chapter 3, Section 2).

2. INTERLIMB REFLEX INFLUENCES DURING REST

It was determined in a study by Sherrington and Laslett (1903) that stimulation of a forelimb nerve generates a response in the ventral roots of the lumbar segments. Later, Lloyd (1942) and Lloyd and McIntyre (1948) made a systematic study of descending intersegmental "forelimb-hindlimb" reflexes evoked by electrodermal stimulation or by stimulation of the muscular or cutaneous nerves in the cat forelimb. It was discovered that the group II, but not the group I, afferents of the forelimbs participate in "forelimb-hindlimb" reflexes. According to Lloyd, interlimb influences are mediated by the descending propriospinal system.

In experiments in anesthetized cats, Buser et al. (1963) found that electrodermal stimulation of the forelimb evokes bilateral responses in the pyramidal tracts, the ventral lumbar roots, and the muscular nerves of the hindlimbs (the latency is approximately 20 msec in the latter case). Although stimulation of the forepaw generates responses in the pyramidal tracts (which are abolished by ablation of the sensorimotor area of the cortex), reflex responses in the ventral lumbar roots are preserved after bilateral pyramidotomy.

It was determined in a study by Abrahams (1970) that destruction of the cortical layer in the cat motosensory area, like decerebration, leads to the sharp depression or abolishment of descending cervicolumbar reflexes and reflexes from forelimb nerves to hindlimb muscles. On the basis of his findings, Abrahams suggests that the ascending cortical pathway for cervicolumbar reflexes is crossed, while the descending pathway from the brain is ambilateral.

Descending interlimb influences apparently include both presynaptic (Schmidt and Willis, 1963; Bergmans and Colle, 1964) and postsynaptic mechanisms governing influences on the interneurons of the spinal cord (Adamovich and Borgest, 1968; Adamovich et al., 1969a,b).

We studied changes in the reflex excitability of the spinal motoneurons of the GM evoked by stimulation of the ulnar nerve (UN). In addition, we studied lower-limb reflex movements provoked by stimulation of the UN. A clinical physiological analysis was performed in an effort to analyze the nature of descending interlimb influences, and the effects of the interaction of vestibulospinal and descending interlimb influences were compared.

Descending Interlimb Reflex Facilitatory Effect

The minimum delay for detection of the "ulnar" facilitatory effect is approximately 35 msec. When the delay is increased further, the "ulnar addition" (UA)--i.e., the difference, expressed as a percentage, between the conditioned and control H-responses--increases smoothly and continuously for 60-70 msec up to a 100- to 110-msec delay (see Fig. 31B). At a delay of 110 msec, the amplitude of the "ulnar" H-response of the GM following strong stimulation of the UN (maximal M-response of m. hypothenar) is, on the average, twice the amplitude of the control ("pure") H-response. Delaying the test stimulation longer than 110 msec after the conditioning stimulation of the UN leads to a gradual decline in the growth of the H-response. After 0.3 sec of stimulation of the UN, only a slight increase in the reflex excitability of the GM motoneurons is retained.

Thus, the time course of the increase in the reflex excitability of GM motoneurons after conditioning stimulation of the ipsilateral UN is similar to that following electrical vestibular stimulation. In both cases, a relatively long latent period is followed by a smooth increase in the reflex excitability of the GM motoneurons for about 60-70 msec. The latency of the "ulnar" effect is about 10 msec longer than that of the "vestibular" effect, and the facilitatory effect achieves a maximum correspondingly 10 msec later after stimulation of the UN than after the onset of vestibular stimulation. As for the second phase, characterized by a gradual decline in reflex excitability to the original level, the "ulnar" effect is reminiscent in this respect of the spinal effect subsequent to vestibular stimulation several milliseconds in duration.

Within limits, the facilitatory effect for the GM motoneurons increases with increasing stimulus strength to the UN (measured for a 100-msec delay). However, during stimulation of the UN that is adequate to evoke an M-response from m. hypothenar with an amplitude about 70% of the maximal, the amplitude of the "ulnar" H-response is, on the average, approximately twice the amplitude of the control H-response, while a further increase in the stimulus strength to the UN has little effect on the magnitude of the UA.

The time course of the contralateral "ulnar" effect is, in principle, similar to the ipsilateral. Their thresholds are also roughly the same (measured for a 100-msec delay), but the magnitude of the threshold contralateral effect is somewhat smaller, and with increasing stimulus strength to the UN above the threshold, the contralateral UA increases more slowly than the ipsilateral. However, at high stimulus strengths to the UN, the contralateral UA does not (on the average) differ from the ipsilateral UA (Mart'yanov, 1968).

Reflex Movements Evoked by Descending Interlimb Reflex Influences

In three subjects, stimulation of the UN induced involuntary contraction (EMG volley) of the GM. For brevity, we shall designate these reflex contractions as "ulnar." The latent period and magnitude of the "ulnar" EMG volley of the GM are to a certain extent dependent on the stimulus strength to the UN. As can be seen from the records in Fig. 32A, the weakest stimulus required for a threshold "ulnar" EMG burst from the GM is subthreshold for the M-response of m. hypothenar. The threshold "ulnar" EMG burst consists of a small number of oscillations (two upper traces in Fig. 32A). With increasing stimulus strength to the UN, the magnitude and duration of the EMG volley of the GM also increase, the volley exhibiting high-amplitude oscillations. This occurs as long as the stimulus strength generates an M-response from m. hypothenar with an amplitude up to 70% of the maximal H-response of this muscle. Increases in stimulus strength beyond this point have no noticeable effect on the "ulnar" EMG volley of the GM.

The greatest duration of the "ulnar" EMG volley does not exceed 50-70 msec, and is thus significantly smaller than the duration of the EMG volley from the same muscles during the performance of a single, rapid voluntary extension of the foot (Fig. 32B). The records in Fig. 32A also show that the increase of the "ulnar" EMG volleys of the ipsi- and contralateral GM parallel the increase in stimulus strength to the UN, while the magnitude of the EMG volley of the contralateral GM is always smaller than that of the ipsilateral GM. At the same time, no significant differences are observed in the thresholds for evocation of the ipsi- and contralateral "ulnar" EMGs.

If the stimulating electrode is moved from the nerve (as indicated by a decrease or by the disappearance of the M-response of m. hypothenar), the "ulnar" EMG volley of the GM does not appear even at the same current strength. Hence, electrodermal stimulation does not evoke "ulnar" movements.

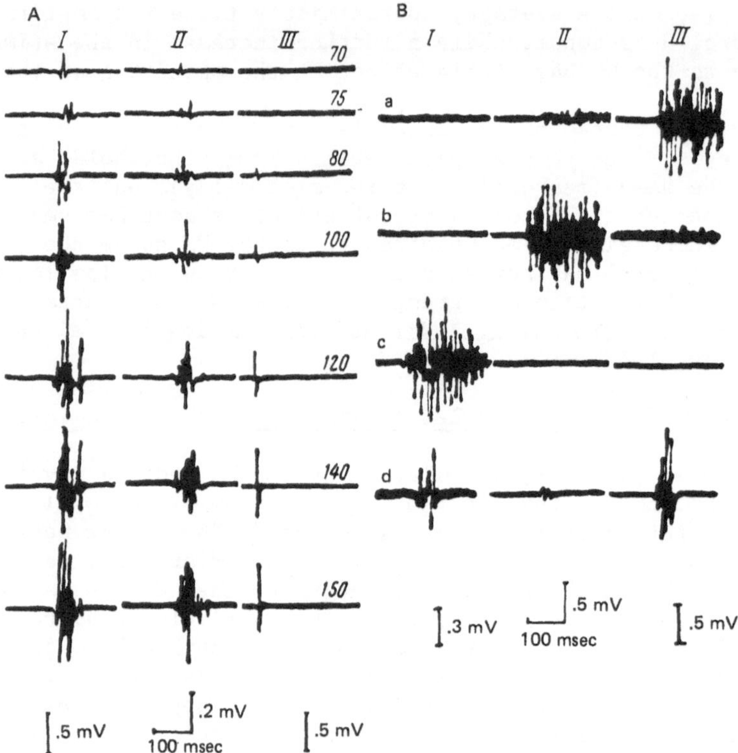

Fig. 32. Reflex contraction of foot extensors evoked by stimulation of the left ulnar nerve. (A) EMGs of left GM (I), right GM (II), and hypothenar muscle of left hand (III). Stimulus strength to nerve increases from top to bottom. (B) EMGs of right GM (I), left ATM (II), and left GM (III); top to bottom: voluntary extension of left foot (a), flexion of left foot (b), extension of right foot (c), and reflex contraction of GM in response to maximal stimulation of left UN (d).

The most prolonged and variable latent period of the "ulnar" EMG volley is observed at threshold stimulus strengths to the UN (from 140 to 180 msec). With a further small increase in stimulus strength, the minimal latent period is shortened to 120-130 msec, and the duration of the latent periods is less variable, although responses with a latency of 140-160 msec, or the complete absence of a response, are still possible. With stimulus strengths to the UN sufficient to evoke an M-response from m. hypothenar approximately 70% of the maximal and stronger, the latency of the "ulnar" EMG volley of the GM is quite constant (about 120 msec), and usually does not vary by more than 10-20 msec. The average duration of the latent period of the "ulnar" EMG volley (based on 20-30 measurements) is as follows:

Subject	Latent period, msec			
Kh.	123.5±0.9	122.0±1.4	126.2±2.8	
Yu.	118.5±1.7	125.0±1.6	117.1±1.6	116.0±1.7
D.	122.0±1.4	127.0±1.8	120.6±0.3	116.7±5.8

The latency of the contralateral "ulnar" EMG volley of the GM at low stimulus strengths to the UN is usually 5-10 msec longer than the latency of the ipsilateral EMG volley of the GM, but at high stimulus strengths, the latent periods of the ipsi- and contralateral "ulnar" responses are equal.

The constancy of the latent period, the bilateral nature of the effect, the brevity of the EMG volley, and the dependence of its magnitude on the stimulus strength to the UN, as well as the absence of the effect of electrodermal stimulation, are indicative of the involuntary character of the origin of "ulnar" GM contraction.

It was demonstrated in special experiments that unlike the voluntary movement, the amplitude of the H-response during the first, early phase of the latent period of an "ulnar" movement (i.e., >60 msec prior to the "ulnar" movement) does not differ from the H-response at rest. Thus, the evocation of an "ulnar" movement is not accompanied by a "pretuning" phase. However, as before a voluntary movement, an increase begins in the reflex excitability of the GM motoneurons 55-60 msec prior to the onset of the "ulnar" EMG volley of this muscle. The amplitude of the test H-response continuously increases, according to the length of the delay between stimulation of the UN and evocation of the H-response (Mart'yanov, 1968).

The onset of the "pretuning" increase in the reflex excitability of the GM motoneurons observed prior to an "ulnar" movement coincides with the onset of the excitability increase of the same motoneurons following stimulation of the UN in cases in which such stimulation does not evoke an "ulnar" movement. The onset of the impulse activity of the GM motoneurons in the case of an "ulnar" movement (120 msec after stimulation of UN) coincides well with the period of maximum motoneuronal reflex excitability induced by stimulation of the UN in the absence of an evoked "ulnar" movement (about 110 msec delay + 10 msec "afferent" time in H-reflex arc).

It is significant that the period of the "tuning" increase in the reflex excitability of the GM motoneurons prior to an "ulnar" movement is of the same duration as that prior to a voluntary movement (55-60 msec). At the same time, the facts presented above demonstrate that these two movements are definitely different in nature.

It is interesting that stimulation of the UN often leads to the appearance of impulse activity by the GM motoneuron pool, while even intense stimulation of the vestibular apparatus, as well as other stimuli (e.g., sound; see Pal'tsev and El'ner, 1967) that also induce significant facilitation of the GM motoneurons, never evoke the impulse activity of these motoneurons. It may thus be supposed that unlike other descending reflex influences, interlimb influences induced by stimulation of the UN have a "triggering" action, similar to voluntary influences. One cannot rule out the possibility that such a similarity between voluntary and "ulnar" influences is due to the presence of some common mechanism by which both types of influence are associated to some degree with the activity of the same supraspinal structures and (or) are mediated over the same pathways. In this regard, it should be recalled that in the cat, descending forelimb-hindlimb reflexes are associated with the participation of the motosensory area of the cortex, and are apparently mediated via both the pyramidal and extrapyramidal pathways (Buser et al., 1963; Abrahams, 1970).

Structures Mediating Descending Interlimb and Vestibulospinal Reflex Influences

To determine the role of the pyramidal system in the origin of descending reflex influences, we did studies in patients with pyramidal paralyses of lower-limb voluntary movements. We examined 8 patients with lower-limb pyramidal paralysis of varying etiology. Of these, 3 cases involved hemiparesis that developed subsequent to the surgical removal of a brain tumor (in the frontal and fronto-parietal areas of the cerebral hemispheres); 3 cases were the result of injury; and 2 involved paraparesis of the lower limbs associated with a tumor in the thoracic cord (with profound motor impairment of the right side in one case, and of the left side in the other).

In every case of pyramidal paralysis (including those with a spinal localization), the facilitatory effect was absent on the affected side on stimulation of the ipsilateral UN, and was severely reduced or totally absent on stimulation of the contralateral UN. At the same time, in pyramidal patients with a cerebral focus, electrical stimulation of the vestibular apparatus (ipsi- or contralateral) evoked distinct vestibulospinal facilitatory effects in the motoneurons of the paralyzed GM. However, in cases involving the complete interruption of the left VIIIth nerve (as described in the previous section), and in the absence of a vestibulospinal facilitatory effect from stimulation of the left ear, stimulation of each UN produced an increase in the reflex excitability of the motoneurons of the ipsi- and contralateral GM.

These facts allow us to conclude that unlike the vestibulospinal facilitatory effect, the descending interlimb reflex facilitatory effect is evidently strongly dependent on the activation of the same central supraspinal (and descending) structures that are necessary for the initiation (triggering) of a voluntary movement. Such a conclusion is in agreement with experimental data obtained in acute experiments in cats (Buser et al., 1963; Abrahams, 1970). It was determined in these experiments that descending interlimb reflexes are abolished or profoundly weakened by destruction of the motosensory cortex. Our data provide evidence that reflex descending interlimb influences in man are mediated largely via the pyramidal pathways. Vestibulospinal effects evoked by electrical stimulation of the vestibular apparatus are mediated via descending pathways distinct from the pathways for interlimb reflexes.

Differences between the vestibulospinal and interlimb descending influences were also discovered by investigating the interaction of these two types of influences. In studying the interaction of unilateral influences, the isolated effects of stimulation of the UN and vestibular apparatus on the same side and the effect of their simultaneous action (change in H-response of GM) were determined in one and the same experiment. It was found that regardless of the polarity of the vestibular stimulus, the ipsi- or contralateral facilitatory spinal effects from the simultaneous stimulation of the UN and vestibular apparatus of the same side actually represent the sum of the two facilitatory effects from each of the stimuli taken separately (A. V. Syrovegin). The results of clinical physiological studies have shown that the vestibular and "ulnar" influences are mediated (if only in part) by different descending spinal pathways. Hence, the results of the joint action of unilateral vestibular and descending interlimb influences may be determined to a certain extent by the interaction of these influences at an interneuronal spinal apparatus common to both pathways and (or) at the spinal motoneurons themselves.

Another method of investigating the interaction of vestibular and descending interlimb influences is to determine the effect of galvanic stimulation of the ipsilateral vestibular apparatus on the "ulnar" movement (Mart'yanov, 1968). To do this, we recorded the "ulnar" EMG volleys of the GM in response to ipsilateral stimulation of the UN and simultaneous stimulation of the UN and vestibular apparatus in subjects in whom such an effect was present. For each of these two cases, the average latency was determined on the basis of 15 measurements. The results of these experiments are summarized in Table 9.

Although significant differences between the latencies for single and combined stimulation were observed only in 4 experiments, in all other experiments, the latency was still shorter after combined stimulation than after stimulation of the UN only.

Table 9. Latency (in msec) of "Ulnar" EMG Volley of GM Evoked by Single Stimulation of Ulnar Nerve (UN) and by Combined Stimulation of UN and Vestibular Apparatus (UN + V)

Experiment No.	Subject Kh.		Subject Yu.		Subject D.	
	UN	UN+V	UN	UN+V	UN	UN+V
1	128.1 ±3.0	123.0 ±3.5	120.1 ±1.9	107.9 ±1.9	130.0 ±3.8	110.9 ±1.1
2	119.0 ±2.2	106.7 ±1.5	112.9 ±2.2	104.0 ±2.2	113.0 ±1.9	108.0 ±3.6
3	126.9 ±2.6	115.8 ±2.0	131.8 ±2.5	121.3 ±4.6	--	--
Average value	124.7 ±2.2	115.1 ±2.8	121.2 ±3.2	111.0 ±2.9	121.5 ±2.6	109.5 ±1.5

According to the data obtained from all the experiments, the average latency of the "ulnar" EMG volley of the GM was significantly reduced by the simultaneous stimulation of the UN and the vestibular apparatus (from 122.5 msec to 111.8 msec).

Subsequent experiments revealed that galvanic vestibular stimulation prior to stimulation of the UN also shortens the latency of the "ulnar" EMG volley of the GM, depending somewhat on the interval between vestibular and ulnar stimulation (Table 10).

In Table 10, the figures given in rows 2 and 3 were obtained by direct experimentation, while the remaining figures were derived by calculation. In rows 2 and 3, the latency was measured as the interval from stimulation of the UN to the onset of the EMG volley of the GM. The figures in row 4 were derived as the difference between the figures in rows 2 and 3; the data in row 5, as the sum of the figures in rows 1 and 3, giving the latency of the EMG volley from the onset of vestibular stimulation.

The figures in row 6 were obtained by the following calculations: As was described earlier, the "tuning" increase in the reflex excitability of the GM motoneurons commences approximately 50 msec after stimulation of the UN. To determine the duration of the "tuning" period, 15 msec was subtracted from the latency of the "ulnar" EMG volley (figures in row 3) for efferent conduction, and

Table 10. Latency (in msec) of "Ulnar" EMG Volley for Simultaneous or Preliminary Galvanic Stimulation of Vestibular Apparatus

Row	Index	Time (msec)				
1	Delay from onset of vest. stim. to stim. of UN	0	40	50	70	100
2	Latency of "ulnar" EMG volley of GM without vest. stim.	122.5	120.7	120.0	119.1	124.0
3	Latency of "ulnar" EMG volley with vest. stim.	111.8	100.0	85.1	67.8	69.0
4	Shortening of latency of "ulnar" EMG volley during vest. stim.	10.7	20.7	34.9	51.3	55.0
5	Latency of EMG volley from onset of vest. stim.	111.8	140.0	135.1	137.8	169.0
6	Approx. duration of "ulnar" facilitation preceding onset of "ulnar" EMG volley	50	35	25	10	10

50 msec for the latency of the "tuning" effect. For example, for a 100-msec latency of the EMG volley (onset of vestibular stimulation 40 msec prior to stimulation of UN--see column 2 of Table 10), motoneuronal impulse activity begins 85 msec after stimulation of the UN. Since the "tuning" changes in motoneuronal excitability begin 50 msec after stimulation of the UN, it follows that the duration of "tuning" in this case is only about 35 msec (see row 6).

The results obtained revealed the unusual character of the interaction of facilitatory vestibulospinal and descending interlimb influences. It was discovered that the reflex EMG volley of the GM arises only after the facilitatory vestibulospinal effect has attained a maximum (see row 5). This is evidenced by the fact that in every case, the EMG volley occurs more than 100 msec after the onset of vestibular stimulation, but at different times after the onset of the "ulnar" facilitatory effect (see row 6). This might well create the impression that the "vestibular" facilitation plays the more essential role, and needs to be augmented by only a slight facilitatory addition (as supplied by "ulnar" facilitation) to evoke a reflex motoneuron volley.

At the same time, however, no impulse activity of the GM motoneurons occurs in these subjects, even if the stimulus strength to the vestibular apparatus is increased, if paired stimulation of the vestibular apparatus is applied with analogous delays, or if an anodal stimulus is applied simultaneously to the two vestibular apparatuses, which leads to the summation of the facilitatory effects in the control (Ya. M. Kots and A. V. Syrovegin). It must be supposed, therefore, that in experiments with preceding stimulation of the vestibular apparatus, stimulation of the UN not only determines the supplementary facilitation of the GM motoneuron pool, which permits the impulse activity threshold to be attained, but also (and most importantly) provides for specific triggering effects on the GM motoneuron pool.

In this regard, the same specific "triggering" role ascribed to voluntary cortical influences should perhaps also be ascribed to descending influences evoked by stimulation of the UN. This role may be due to the fact that descending interlimb influences evoked by stimulation of the UN are mediated by the same "specific" descending triggering pathways (pyramidal, perhaps) that are utilized for the cortical initiation of voluntary movement.

3. DESCENDING REFLEX INFLUENCES DURING THE ORGANIZATION OF VOLUNTARY MOVEMENT

The question regarding the mode and mechanisms of the interaction of different supraspinal motor systems during the organization and regulation of voluntary movement is one of a number of fundamental questions dealing with the physiology of central voluntary motor control. Indeed, as experimental and clinical data demonstrate, motor centers are located at different levels of the brain, and each level has independent descending connections with the spinal motor centers. The possibility of the participation of these supraspinal motor systems in the control of the spinal motor apparatus is confirmed by the presence of motor effects during their stimulation or motor defects after their destruction.

The most general hypothesis regarding the principles of the interaction of different reflex pathways was formulated by Sherrington (1906) in his theory on a final common pathway. According to Sherrington, the final common pathway is a passive tool in the hands of each group of reflex pathways. The changes that occur at each stage of a well-ordered sequence of reactions and that normally shape animal behavior consist essentially of the switching of the final common pathway from one system of reflexes to another. In a scheme such as this, one might expect that the interaction of different supraspinal centers would lead to the dominance of the effects of one system and the suppression of the effects of the other.

In studying this problem, it is particularly important to investigate the interaction between the cortico- and vestibulospinal motor systems, which are known to have different descending pathways and different areas of termination in the spinal cord (Kuypers, 1964). It is significant that both these systems possess direct connections with the spinal motoneurons (see Chapter 3, Section 2). Differences in the functional significance of these two systems are indicative of their continuous interaction. For example, Gernandt and Gilman (1960) point out that as phasic pyramidal activity changes the position of the body, it also induces vestibular stimulation, while the postural changes that evoke the stimulation of the vestibular apparatus must in turn apply their corrections to the phasic movement accomplished with the participation of the pyramidal system.

In experiments with anesthetized and decerebrated cats, Gernandt and Gilman applied the paired-stimulus method of testing--conditioning stimulation of the vestibular nerve and test stimulation of the motor cortex, or vice versa--and recorded the impulse activity of the motoneuronal pool from the radial nerve or from the lumbar ventral roots in order to study the summated effects of the interaction between the vestibular and cortical (pyramidal and extrapyramidal) systems. Stimulation of the cortex and vestibular nerve with recording at the level of the brain stem, as well as the same paired stimulation in pyramidotomized animals with recording at the spinal level, were employed in studying the interaction of the vestibular and cortical extrapyramidal motor systems at various levels; stimulation of the pyramidal tract below the level of its transection and of the vestibular apparatus was used to study the interaction of the pyramidal and vestibular systems at the level of the spinal cord.

The experiments showed that the vestibular and cortical motor systems may interact at various levels; under the given investigative conditions, the vestibular effects were dominant over the cortically evoked motor effects. According to the conclusion of Gernandt and Gilman, in the "battle" between the vestibular and cortical systems for access to the final common pathway, the vestibular-evoked activity is dominant. According to the data of Erulkar et al. (1966), the interaction between the cortico- and vestibulospinal systems may be accomplished at the spinal level by the participation of its interneuronal apparatus. That stimulation of the vestibular nerve can give rise to a volley in the pyramidal tract (Megirian and Troth, 1964) demonstrates that the interaction of these two systems is possible at the cortical level also.

Interaction of Vestibulospinal and Voluntary Influences During the Organization of Voluntary Movement

The unique character of the interaction between these two influences has been demonstrated in experiments designed to determine the latent period of a voluntary movement against a background of galvanic vestibular stimulation. Since the latter produces a substantial and long-lasting increase in the reflex excitability of GM motoneurons (see Section 1 above), it may be supposed that under these conditions, there will be a shortening of the latent period of voluntary foot extension (a movement involving participation of the GM) due to the summation of vestibular and "voluntary" facilitatory effects on the motoneurons of this muscle.

In one series of experiments, the subjects performed an extension of the foot in response to two movement signals that were alternated in random order: electrodermal stimulation of the ipsilateral earlobe and galvanic stimulation of the ipsilateral vestibular apparatus. Both prior to and following the main part of the experiment, we made certain that electrodermal stimulation of the earlobe did not, by itself, produce an increase in the test H-response (see Section 1 above), while galvanic stimulation of the vestibular apparatus increased the test H-response of the GM by approximately 100% (in both cases, the delay in evoking the H-response was 100 msec).

In another series of experiments, the movement signal was an electrodermal stimulation of the earlobe, preceded by 100 msec of either galvanic stimulation of the vestibular apparatus or electrodermal stimulation of the earlobe. The subject was instructed to extend the ipsilateral foot in response to the second signal. The duration of the latent period of the voluntary movement was determined with an accuracy of ±5 msec by the interval from the movement signal (from the single stimulus in the first series of experiments and from the second stimulus in the second series) to the onset of the EMG volley of the GM. The duration of the latent period was determined as the average of 20 measurements of the latent period for each of the signals.

The results of the experiments showed that the average time and variability of a motor reaction in response only to electrodermal stimulation of the earlobe or only to galvanic vestibular stimulation do not differ (122.4 and 122.5 msec, 16.1 and 15%, respectively). Galvanic stimulation of the vestibular apparatus begun 100 msec prior to the electrodermal movement signal does not shorten the latent period of voluntary movement. The average time and variability of a motor reaction in response to an electrodermal stimulation preceded by a galvanic stimulation of the vestibular apparatus do not differ from the time and variability of a motor reaction in response to repetitive electrodermal stimulation: 130.9 and 132.3 msec, 13.5 and 15.6%, respectively.

Thus, on the one hand, conditioning stimulation of the vestibular apparatus produces a significant increase in the reflex excitability of the GM motoneuron pool in conditions of rest; on the other hand, the same vestibular stimulation preceding the movement signal leads to no shortening of the latency of the voluntary impulse activity of this pool. The experiments described below are designed to identify the causes of this phenomenon (Kots and Mart'yanov, 1968; Kots, 1971).

In one series of experiments, the magnitude of the vestibulospinal facilitatory effect was determined during the latent period preceding the onset of voluntary foot extension. In these experiments, a neon light flash served as the movement signal. A "control" test H-response or a "vestibular" H-response was evoked in different periods prior to the onset of voluntary movement; i.e., in the latter case, 100 msec before evocation of the test H-reflex, galvanic stimulation of the ipsilateral vestibular apparatus was begun. One "vestibular" H-response was evoked for every 2-4 control H-responses during the course of the experiment.

The average results of 23 experiments are shown graphically in Fig. 33A. It can be seen that the last 60 msec of the latent period of voluntary movement are characterized by a strong increase in the amplitude of the "control" and "vestibular" H-responses. The amplitude of the "vestibular" H-response is greater than that of the "control" H-response for all but the last 40 msec of the latent period. In the interval 60-35 msec prior to movement, the amplitude of the "vestibular" H-response increases more than the amplitude of the "control" H-response (see slopes of curves 1 and 2). During the last 30 msec of the latent period, the amplitude of the "vestibular" H-response either exhibits no change, or even declines somewhat, such that the amplitudes of the "vestibular" and "control" H-responses are equal during the last 20-msec interval.

Accordingly, during the early ("pretuning") phase of the latent period, the VA does not change and is roughly equal to the VA at rest. (Since the "control" H-response during the first phase of the latent period is greater than when at rest, it follows that the absolute increase in the amplitude of the test H-response from vestibular stimulation in this phase is greater than when at rest.) In the first half of the "tuning" phase of the latent period (in the interval from approximately 60 to 35 msec prior to the voluntary EMG volley), the VA is significantly greater than in the first phase of the latent period or when at rest. During the second half of the "tuning" phase (last 30-msec interval prior to voluntary movement), the VA is sharply depressed and finally disappears.

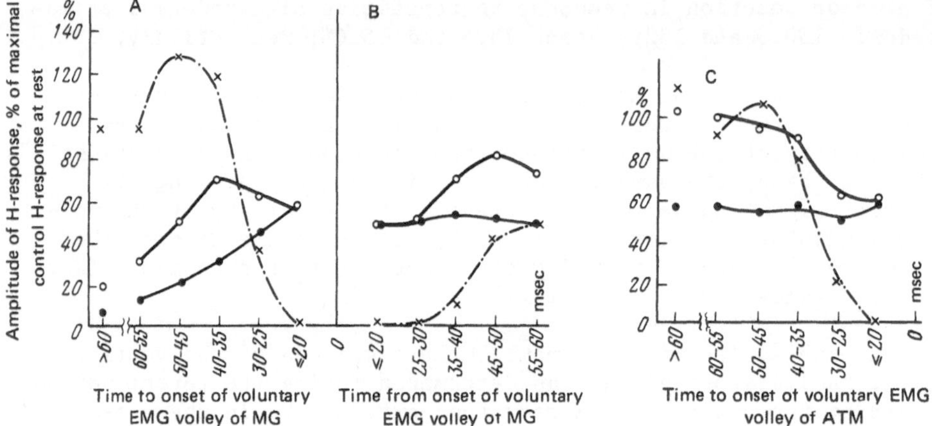

Fig. 33. Changes in vestibulospinal facilitatory effect during course of latent period (A) and at onset of voluntary foot extension (B) and during course of latent period of voluntary foot flexion (C). (•) "Control" H-response of Sol; (o) "vestibular" H-response of Sol; (x) "vestibular addition."

The decrement and disappearance of the VA during the last 30 msec of the latent period of voluntary movement cannot be attributed entirely to the occlusion of vestibulospinal and "voluntary" influences, whereby the "vestibular" component of the facilitatory effect is gradually masked by the more powerful "voluntary" facilitation, but must be at least partly associated with a blockade of evoked vestibulospinal facilitatory influences. First, in 7 of 23 experiments, a decline was recorded in the amplitude of the "vestibular" H-response within the last 30 msec of the latent period relative to the amplitude of the "vestibular" H-response 35-40 msec prior to the onset of voluntary movement. In the other experiments, the "vestibular" H-response was no smaller during the last 20 msec of the latent period than the "vestibular" response recorded 25-30 or 35-40 msec prior to the onset of movement, but in no experiment was the average amplitude of the "vestibular" H-response during the last 20 msec of the latent period significantly greater than the average amplitude of the "vestibular" response 35-40 msec prior to voluntary movement. And in no experiment did the average amplitude of the "vestibular" H-response recorded during the last 20 msec of the latent period exceed the average amplitude of the "control" H-response in the same interval.

In a special series of experiments, the latent period of voluntary movement was tested only by means of the "vestibular" H-response, which made it possible to obtain curves describing changes

Table 11. Average Amplitudes of "Vestibular" H-Response During Latent Period of Voluntary Foot Extension (in % of Maximal Amplitude of H-Response at Rest)

Intervals from H-response to onset of EMG volley (msec)							
More than 60		60-46		45-35		30 and less	
M±m	N	M±m	N	M±m	N	M±m	N
20.8±1.3	9	29.2±1.5	22	42.3±3.5	21	33.1±2.2	21
14.5±1.3	10	20.7±2.7	20	37.0±3.3	20	24.0±2.6	18
41.4±3.6	11	69.2±4.4	18	80.5±4.2	10	68.1±2.8	28
43.5±5.0	17	50.9±4.5	22	61.8±2.7	23	47.9±3.9	19
38.1±6.2	5	51.0±3.5	9	65.2±3.9	14	49.3±4.1	14

in the amplitude of the "vestibular" H-response on the basis of a relatively large number of responses, and to perform a statistical analysis on the basis of small samples. The results of 5 such experiments are shown in Table 11. It can be seen that in all cases, the average amplitude of the "vestibular" H-response was significantly smaller during the last 30 msec of the latent period than 45-35 msec prior to voluntary movement.

All these results permit us to conclude that the decrease in the VA over the last 30 msec of the latent period is a reflection of a true decline in the vestibulospinal facilitatory effect, and is not merely the result of a masking of the vestibular effect due to the voluntary facilitation of the agonist motoneuron pool ("saturation") or to the occlusion of "voluntary" and vestibulospinal influences.

The dynamics of the vestibulospinal facilitatory effect during voluntary foot extension are quite unique (Fig. 33B). Since the "control" H-response of the GM undergoes no appreciable changes during the period of voluntary GM activity investigated, the changes in the "vestibular" H-response in this period reflect changes in the character of evoked vestibulospinal influences. No facilitatory vestibulospinal effect is observed during the first 30 msec of the voluntary EMG volley of the GM agonist.

During later intervals following the onset of the voluntary EMG volley, the vestibulofacilitatory effect increases rapidly. For example, 35-40 msec after the onset of the agonist EMG volley, the VA amounts to only 14.6% (and was entirely absent in 6 of the 16 experiments), while during the next 30-40 msec, it increases to approximately 30% (and was observed in this interval in all 16 experiments). Judging by these data, the period of absence of vestibulospinal facilitatory influences includes the first 30 msec of the voluntary impulse activity of the agonist motoneuron pool.

Thus, the period of absence of evoked vestibulospinal facilitatory effects on the agonist motoneuron pool lasts about 60 msec and includes the last 25- to 30-msec interval prior to the onset of its voluntary activity and the first 30 or so msec of this activity.

The blockade of vestibulospinal facilitatory influences during the latent period of a voluntary movement occurs not only with respect to the motoneuron pool of the future agonist, but also with respect to the motoneuron pool of the future antagonist of the movement. Such a conclusion is supported by the results of experiments involving the "control" and "vestibular" H-reflex testing of the GM prior to voluntary foot flexion, a movement in which the GM assumes the role of direct antagonist. As can be seen from the graph in Fig. 33C, during all but the last 40 msec or so of the latent period preceding voluntary foot flexion, the VA undergoes no noticeable changes. During the last 40 msec of the latent period, however, the VA begins a slow decrease, which accelerates during the last 30 msec and is particularly pronounced during the last 20 msec.

To determine the spinal "topography" of the site of the blockade, it was important to study the vestibulospinal facilitatory effects during the latent period of voluntary movements in other joints. Figure 34 gives portions of records from an experiment involving "control" (A) and "vestibular" (B) H-response testing during the latent period preceding voluntary flexion in the hip joint. During rest, the subject's foot was flat on the floor and thus had to be raised to perform the movement. With such a mode of flexion in the hip joint, weak EMG activity of the GM was recorded along with the EMG volley of the m. rectus femoris--the primary agonists of the movement.

During the course of the latent period of hip flexion, the control H-response of the GM does not change noticeably (see Fig. 34A, traces 2-5), but increases sharply at the onset of the EMG volley of the primary agonist (see Fig. 34A, trace 6). During the course of the entire latent period, the amplitude of the "vestibular" H-response continuously exceeds the amplitude of the control response (see Fig. 34B, traces 2-5). Analogous results were obtained by a determination of the magnitude of the vestibulospinal facilitatory effect during the latent period of voluntary extension of the contralateral (relative to tested leg) foot or

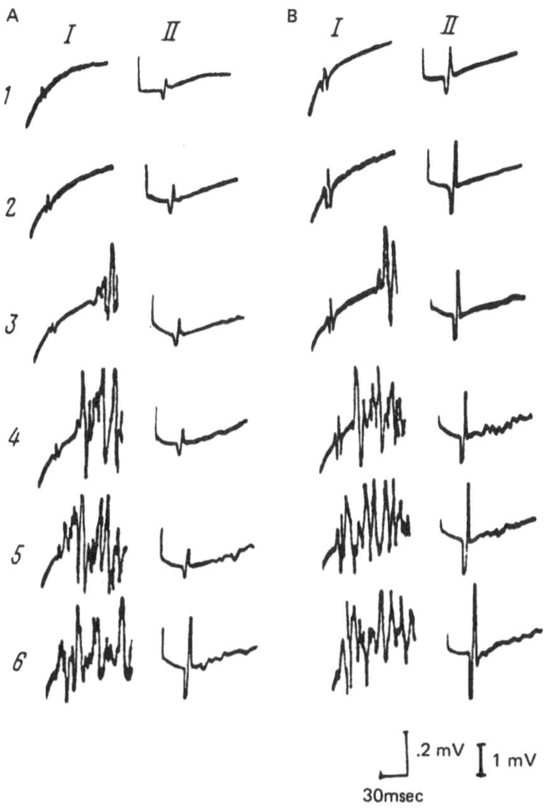

Fig. 34. Vestibulospinal facilitatory effect during latent period of voluntary flexion in the hip joint. (A) Changes in "control" H-response of Sol; (B) changes in "vestibular" H-response of Sol according to proximity to onset (2-5) and at onset (6) of voluntary EMG volley of rectus femoris muscle. (I) EMG of rectus femoris muscle; (II) EMG of Sol. (1) "Control" (A) and "vestibular" (B) H-responses during rest.

voluntary flexion in the ipsilateral elbow joint. It was determined that there are also no changes in the VA prior to these movements.

Thus, the blockade ("switch-off") of vestibulospinal facilitatory influences during the organization of voluntary movement has a strictly local character, since it is observed only with respect to the "active" segments, including the motoneuron pool of the future agonists and antagonists of an active joint. There remains, however, the possibility of the transmission of evoked vestibulospinal facilitatory actions to the other "inactive" segments, including the motoneuron pools of muscles not involved in the voluntary movement.

Interaction of Vestibulospinal and Descending Interlimb Influences During the Latent Period of Voluntary Movement

It was shown in the preceding section that in all those subjects in whom stimulation of the UN evokes the reflex contraction of the GM ("ulnar" movement), galvanic stimulation of the vestibular apparatus begun 100 msec prior to stimulation of the UN leads to an appreciable curtailment of the latency of an "ulnar" movement (by about 50 msec on the average). These experiments were repeated with one difference: evocation of the "ulnar" movement against a background of galvanic stimulation of the ipsilateral vestibular apparatus was alternated between evocation at rest (variant 1) and in circumstances in which UN stimulation (also against background galvanic stimulation of the vestibular apparatus) served as the signal for voluntary extension of the ipsilateral foot (variant 2). The latency of the "ulnar" movement for each variant was calculated on the basis of 20 movements. The data obtained for three subjects are as follows:

Subject	Latent period, msec	
	Variant 1	Variant 2
Kh.	73.1±3.6	115.5±1.8
Yu.	64.4±4.3	117.0±2.1
D.	69.0±5.2	110.0±3.2

Thus, in conditions of rest, galvanic stimulation of the vestibular apparatus leads to a considerable curtailment of the latency of a reflex "ulnar" movement, while during the latent period of a voluntary movement, the same stimulation of the vestibular apparatus produces no changes in the latency of the "ulnar" movement. Such a result may be regarded as yet another manifestation of the blockade of vestibulospinal facilitatory influences on the agonist motoneuron pool prior to voluntary movement. In the present case, however, it is unclear whether only the vestibulospinal influences are blocked, or also those influences that lead to the occurrence of the "ulnar" movement. Nevertheless, the results of the experiments described below seem to exclude the second possibility.

The functional significance of the processes discussed above seems to be clear. The supraspinal motor centers (most likely the cortical) the descending command of which provides for the "triggering" of a voluntary phasic movement also provide for the blockade ("switch-off") of vestibular, and perhaps other (see Korobkov et al., 1969; Zalkind, 1972), direct actions on the "active" segments of the spinal cord. This excludes the possibility of "extrinsic" actions on these segments, and the process of their control is transformed into the exclusive "affair" of the voluntary motor centers. It should be emphasized that we are speaking only of the temporary suspension of descending vestibulospinal (and perhaps other "involuntary") influences. At the same time, it is quite probable that during this blockade ("switch-off") period, afferentation from the vestibular apparatus continues to reach the voluntary cortical centers by way of ascending vestibulocortical connections (Megirian and Troth, 1964), where it is utilized in the formulation of the supraspinal command.

During the "switch-off" period, the spinal motor centers of those muscles that are not the direct objects of the authoritarian cortical control probably remain accessible for direct actions from the vestibulospinal system, which enables this system to rapidly supply direct corrections for the avoidance and restoration of postural disruptions associated with the performance of a phasic voluntary movement.

We still do not know at which levels of the CNS the interaction of voluntary (cortical) and vestibulospinal influences occurs. Judging by experimental data, such interaction is possible at various levels (Gernandt and Gilman, 1960; Megirian and Troth, 1964; Erulkar et al., 1966), although considering the fine topography of the blocking effect, we believe it is not likely the result of spinal changes evoked by supraspinal (cortical) command.

Descending Interlimb Reflex Influences During the Organization of Voluntary Movement

An investigation of the action of descending interlimb influences during the organization of voluntary movement was performed in subjects in whom a reflex "ulnar" movement (contraction of GM) was evoked by stimulation of the UN. When UN stimulation was transformed into the signal to perform a voluntary foot extension, the latency of the voluntary EMG volley of the ipsilateral GM was of the same order as that of the "ulnar" reflex volley at rest (Fig. 35, column 1, cf. A and B). Disregarding rare cases in which the latent period of the voluntary movement differs appreciably from the latency of the "ulnar" movement, we may say that the average duration and variability of the latency of the voluntary and "ulnar" EMG volleys of the GM coincide.

Due to this coincidence, it is quite difficult to determine the nature of the initial part of the EMG volley of the GM during the performance of a voluntary movement in response to stimulation of the UN (movement signal). It may be the reflex ("ulnar") EMG volley evoked by stimulation of the UN, followed by the "volun-

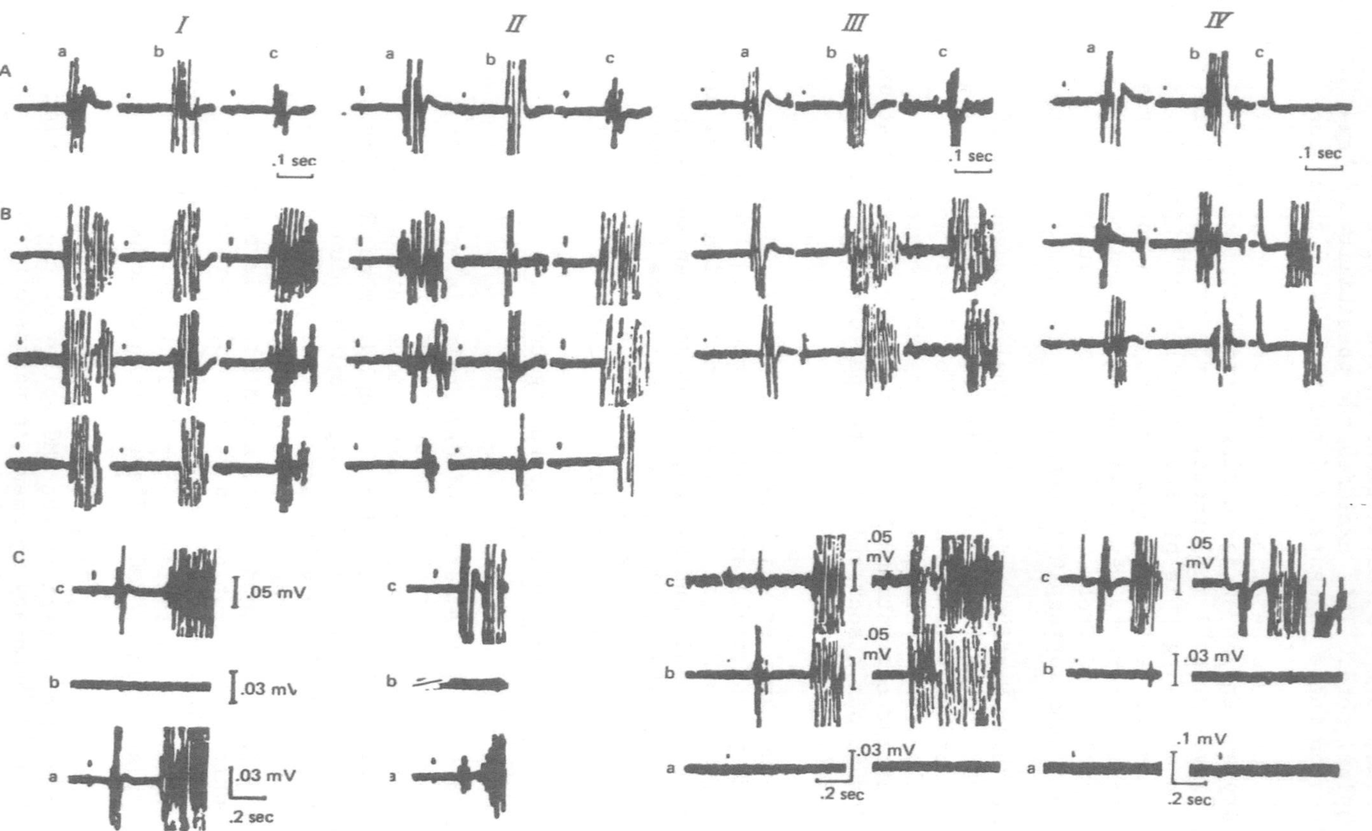

Fig. 35. Voluntary movements in response to stimulation of the left UN: extension of left foot (I), flexion of left foot (II), extension of right foot (III), and flexion of left elbow joint (IV). (A) Nerve stimulation during rest; (B) nerve stimulation--signal to move with no delay; (C) nerve stimulation--signal to move with delay. (a) EMG of left Sol; (b) EMG of right Sol; (c) EMG of left ATM (I and II), right ATM (III), and left biceps brachii (IV).

tary" activity or the voluntary EMG activity the relatively brief and low-variability latency of which is associated with "ulnar" facilitation, that serves as the triggering signal for the voluntary impulse activity of the motoneurons.

It is of some interest that during the performance of foot extension in response to UN stimulation, the EMG volley of the GM is much longer than the EMG volley of the GM during a purely reflex "ulnar" movement, and it is often possible to distinguish in it a leading volley less than 100 msec in length, followed, after a very short, sharp decline or even the absence of activity, by a volley of much greater duration (see Fig. 35I, B, middle trace). This circumstance favors the first hypothesis, whereby the initial EMG volley of the GM reflects the enhanced reflex "ulnar" activity, while the subsequent EMG activity is "voluntary" in nature. In both cases, it must be assumed that during the latent period of voluntary movement, there is no blockade of evoked descending interlimb influences on the agonist motoneuron pool.

Such an assumption is also supported by the observation of an EMG volley from the contralateral GM, which is recorded during the performance of voluntary extension of the ipsilateral foot in response to stimulation of the UN (see Fig. 35I, B). It is characteristic of this volley that (1) its duration is of the same order (50-70 msec) as in conditions of rest and (2) its onset never leads, and usually coincides with, the onset of the EMG volley of the primary agonist of the voluntary movement--the ipsilateral GM. During a voluntary movement made in response to any other signal (audible or visual) or without a signal (by the subject's own initiative), no EMG activity is recorded from the symmetrical muscles of the other limb.

These facts permit us to regard the EMG volley of the contralateral GM during voluntary extension of the ipsilateral foot in response to UN stimulation as a reflex "ulnar" EMG volley retained under these conditions. It is significant that the time of the appearance of the contralateral "ulnar" volley is tied to the onset of the voluntary agonist EMG volley of the ipsilateral limb. This is seen in cases in which the latency of voluntary extension of the ipsilateral foot is clearly greater than the usual latency of an "ulnar" movement (see Fig. 35B, column I, lower trace). The close connection between the onset of the "ulnar" EMG volley of the GM and the onset of voluntary agonist activity is also observed during the performance of other voluntary movements in response to stimulation of the UN: flexion of the ipsilateral foot (column II), extension of the contralateral foot (column III), or flexion of the contralateral foot, ipsilateral hip, or ulnar (column IV) flexion.

These data permit us to conclude that during the performance of a voluntary movement in response to stimulation of the UN, there is no blockade of descending interlimb influences that evoke the bilateral reflex contraction of the GM in response to UN stimulation. The short latency and bilateralness of the effect indicate that the brief EMG volleys of the GM recorded during the performance of voluntary movements in response to stimulation of the UN are of the same nature as the reflex "ulnar" EMG volleys of these muscles at rest.

At the same time, the onset of the reflex "ulnar" EMG volley is rigidly tied to that of the voluntary EMG volley. In particular, the "ulnar" activity of the GM motoneurons never begins earlier than the voluntary activity of the other motoneuron pools of the agonists, and usually (if the latency of the voluntary activity is no shorter than 110 msec) these two forms of activity commence simultaneously. These facts suggest a definite interaction of "voluntary" and reflex "evoked" activity. The impression is created that during the organization of voluntary movement, the higher voluntary motor centers have control of descending interlimb influences on the motoneuron pools of muscles not involved in the voluntary movement. The results of experiments described below also indicate that such control pertains as well to descending reflex influences on the motoneuron pools of the future agonists.

In the experiments, the subjects (in whom "ulnar" movements were recorded during rest) performed various movements in response to stimulation of the UN, with the onset of voluntary movement deliberately delayed (by 0.25-0.8 sec). The following movements were performed: extension or flexion of the ipsilateral (relative to stimulated UN) or contralateral foot, flexion in the ipsilateral hip joint, flexion in the ipsi- or contralateral elbow joint. In most cases, during the course of the entire latent period of any voluntary movement performed with a deliberate delay in response to stimulation of the UN, no electrical activity of the future agonist of the movement is observed, nor that of any other muscle, including both GMs. After several repetitions of the movement, however, a brief (usually 50- to 80-msec) EMG volley is found to arise in the future agonist of the voluntary movement prior to the onset of the basic "voluntary" activity of this muscle, and is separated from the last distinct silent period (see traces in Fig. 35C). This EMG volley preceding voluntary movement has the following characteristcs:

1. It arises only when stimulation of the UN serves as the movement signal, and is absent whenever other signals are given (light flash, audible click) that during rest produce no changes in the electromyographic activity of the GM or the other muscles of the subjects in question. We therefore designate such an EMG volley as "ulnar."

2. It arises only in the future agonist of the voluntary movement, regardless whether the "ulnar" EMG volley is recorded in this muscle during rest. Thus, the "ulnar" EMG volley is recorded in the ipsilateral GM during the prolonged latent period of voluntary extension of the ipsilateral foot (see Fig. 35C, column I) or in the contralateral GM prior to the onset of voluntary extension of the other foot (see Fig. 35C, column III). In these muscles, however, the "ulnar" EMG volley is also recorded at rest. But an "ulnar" EMG volley is recorded in the ipsi- or contralateral ATM during the delayed latent period of voluntary flexion of the ipsi- (see Fig. 35C, column II) or contralateral foot, respectively. This volley is recorded in the flexors of the hip joint during the prolonged latent period of flexion in this joint, and in the ipsi- or contralateral biceps branchii prior to flexion in the elbow joint (see Fig. 35C, column IV). In none of these muscles does stimulation of the UN during rest evoke an "ulnar" EMG volley.

3. The "ulnar" EMG volleys are absent in all muscles not involved in the voluntary movement, including the GM, in which "ulnar" EMG volleys are recorded during rest. (The weak electrical activity of the ATM [antagonist] accompanying the "ulnar" EMG volley of the GM or the weak electrical activity of the GM [antagonist] accompanying the "ulnar" EMG volley are of the "concomitant" type--see Chapter 4, Section 2.)

4. The latency of the "ulnar" EMG volley during a delayed voluntary movement is approximately 120 msec for the muscles of the ankle joint--the GM and ATM--about 110 msec for the femoral muscles, and about 90 msec for biceps brachii, and it is quite constant during the repetitive generation of this volley. Differences in the latency of the "ulnar" EMG volley of these muscles are adequately accounted for by differences in the conduction times to the muscles (Blinkov et al., 1968). It is significant that the latency of the "ulnar" EMG volley of the GM during the delayed latent period of the voluntary movement is equal to the latency of the reflex "ulnar" EMG volley of this muscle at rest, which provides still more evidence in favor of the hypothesis concerning the common nature of these two volleys.

The observations of Luschei et al. (1967) are similar in several respects to the phenomenon described. These investigators noted the appearance of a brief EMG burst in the future agonist during the latent period of a conditioned reflex movement by monkeys some 25-50 msec after presentation of the movement signal. The latency of this burst did not change after the animals were trained to delay the onset of movement.

It is remarkable that the latency of the "ulnar" EMG volley of muscles other than the GM during the prolonged latent period of voluntary movement coincides with the duration of the most frequent latent period of a voluntary movement performed in response to stimulation of the UN with no delay. These results give us reason to believe that the onset of agonist activity during the performance of a voluntary movement in response to stimulation of the UN with no delay is associated with descending interlimb reflex influences on the motoneuron pool of the agonist of the voluntary movement.

The present series of experiments has revealed yet another physiological manifestation of the central "pretuning," i.e., the readiness to perform a given voluntary movement. This "pretuning" state is manifested in the present case by a phenomenon strongly reminiscent of A. A. Ukhtomskii's phenomenon of dominance (Ukhtomskii, 1950; Vasilevskii, 1968; Rusinov, 1969): afferent influences that are indifferent to a given motoneuron pool during rest evoke the impulse activity of this pool during the period of readiness to perform a voluntary movement if the pool is to play the active role of agonist in the future movement. Conversely, if the motoneuron pool is not to participate in the voluntary movement, the afferent influences that evoke its impulse activity at rest evoke no activity during the "pretuning" phase of the movement.

One of the mechanisms that promote the initiation of the impulse activity of the motoneuron pool of the future agonist of the voluntary movement in response to stimulation of the UN is probably the "pretuning" increase in the reflex excitability of the agonist motoneuron pool (see Chapter 3, Section 2). We are inclined to believe, however, that the latter phenomenon is itself a spinal reflection of central events occurring at higher levels of the voluntary motor control system during the organization of voluntary movements. Evidently it is these events that are most important in determining the initiation of reflex "ulnar" impulse activity in those motoneuron pools the muscles of which are to assume the role of primary agonists in the voluntary movement.

CONCLUSION

At the higher levels of the CNS responsible for the regulation of motor behavior, the process of the organization of voluntary movement includes the formulation of a general motor problem determined by concrete conditions and the goals of the individual, as well as the choice of the most adequate forms of motor behavior for solving the motor problem at hand (Bernstein, 1947, 1966; Luriya, 1962, 1970), the supraspinal command being formulated in accordance with this choice. At present, almost nothing is known about the structures that realize these processes. The neurophysiological study of the mechanisms governing the organization of voluntary movement has virtually just begun. The functional significance of a number of physiological processes that are recorded in various supraspinal structures prior to voluntary movement remains obscure. It is certain, however, that the complex process of the central (supraspinal) organization of voluntary movement requires a considerable amount of time.

If the great complexity and long duration of processes in the higher supraspinal regions of the motor system that determine a voluntary movement were regarded as "natural" by the physiologist, the existence of long and complex changes in the spinal neuronal apparatus that precede the onset of movement was not so obvious *a priori*. However, our studies have demonstrated the existence of long and complex changes in the spinal motor apparatus preceding the onset of voluntary movement. These changes are determined by supraspinal influences--the supraspinal command for a voluntary movement--and are considered by us to be the spinal component of the complex process of the central organization of voluntary movement.

The existence of complex and prolonged spinal changes preceding the onset of voluntary movement is in agreement with the hypothesis regarding a hierarchical system for the control of movement (Jackson, 1873; Bernstein, 1947), in which each underlying level is not merely a slave relay for the undistorted transmission of signals originating at a higher level, but is in large measure an active integrative component of the system as a whole (Gel'fand et al., 1961, 1962, 1966; Gel'fand and Tsetlin, 1962, 1966; Gurfinkel' et al., 1965; Alekseev, 1967; Rokotova et al., 1969, 1971). This concept of the functional organization of the CNS was developed by Soviet neurophysiologists, and is beginning to gain wide acceptance among researchers abroad. As A. A. Ukhtomskii wrote: Many peripheral instruments of neural regulation are not simply the mediators of regulatory actions from the cerebrum, or the implementors of a fixed number of 'subcortical stereotypes'.... The cortex must frequently proceed from regulatory actions engendered and summed at lower levels...." (Ukhtomskii, 1954, p. 78). With respect to the regulation of motor function, L. A. Orbeli explained: "There are no discrete centers of coordination; coordination is the chief function of the central nervous system, of every section of the central system. Each part of the central nervous system provides for the transmission of excitation together with a particular form and degree of coordination" (Orbeli, 1934, p. 49).

In a survey article in the Handbook of Physiology, Paillard (1960) writes: "Thus, the spinal keyboard does not receive the commands from higher levels as a docile instrument ready to transmit, blindly and faithfully, carefully prepared orders to the muscles. It appears as a fine machinery sensitive to certain types of influences, influenced by messages of diverse origins which, directly or indirectly by way of the internuncial systems, converge upon it. It already constitutes in itself an 'integrative' structure in the sense that Sherrington gave to this term" (p. 1686).

Indeed, as our findings indicate, the supraspinally evoked changes in the state of the spinal neuronal apparatus prior to voluntary movement cannot be reduced solely to the processes of the supraspinal facilitation of agonist motoneurons that directly provide for their impulse activity. An essential feature of the spinal processes observed during this period includes changes in the state of various interneuronal systems linked with the motoneuron pools of the agonists (see Chapter 3, Section 4) and antagonists (see Chapter 4, Section 3) of a future voluntary movement. We thus have reason to conclude that the change in the state of the interneuronal apparatus of the spinal cord is an important mechanism of the organization of voluntary movement.

The character and degree of changes in a particular spinal interneuronal system are probably determined by the concrete properties of the impending movement. For example, in a study of the state of the interneuronal spinal reciprocal inhibitory apparatus acting on the antagonist motoneuron pool, it was determined (see Chapter 4, Section 3) that training leads to an enhancement of the premotion supraspinal "tuning" of the SRIA, which in turn enhances the reciprocal inhibitory actions on the antagonist motoneuron pool during the performance of the movement, thereby ensuring a more efficient interaction of antagonist muscles.

Another example of the specificity of the supraspinal regulation of the interneuronal spinal apparatus that is determined by the character of the motor activity is the absence of the "silent period" (following the reflex discharge of motoneurons) immediately prior to the onset, or at the onset, of a voluntary phasic movement, as opposed to the presence of this "silent period" during a voluntary static effort. An analysis of this phenomenon as well as other data (see Chapter 3, Section 4) lead to the conclusion that the organization of a voluntary movement is evidently accompanied by the inhibition of those spinal inhibitory interneuronal systems (probably including the Renshaw cells) that are responsible for the origin of the "silent period." This permits a high motoneuronal discharge frequency and the near-synchronous firing of a large number of agonist motoneurons, which is necessary for the performance of this type of movement. In the case of a voluntary static effort, the spinal inhibitory interneuronal system responsible for the "silent period" remains active, thereby ensuring a different mode of motoneuronal response that is characteristic of static muscle tension: a relatively low firing frequency and asynchronous impulse motoneuron activity (Gel'fand et al., 1963).

In this respect, our findings are in agreement with data obtained in acute experiments in animals, according to which all descending motor systems establish primary connections with different interneuronal systems of the spinal cord (see Chapter 3, Section 4, and Chapter 4, Section 3). It is very likely that the supraspinal levels of motor control, by their action on the segmental interneuronal systems, assign that system of segmental neuronal interaction that is necessary for the realization of a required movement (Pyatetskii-Shapiro and Shik, 1964; Gurfinkel' et al., 1965; Feldman, 1966). Direct supraspinal action on the motoneurons via monosynaptic connections or by a special "intercalated" spinal interneuronal system selectively activated by supraspinal influences may be only a part of this process (see Chapter 3, Section 2).

It is quite probable that the mode of the control of segmental motor levels by higher motor levels is merely a reflection of a general mode of interaction that exists between the different levels of the hierarchical motor control system, as was postulated in a model put forward in a study on the multilevel system of motor control (Pyatetskii-Shapiro and Shik, 1964) based on the principle of "minimal interaction" of Gel'fand and Tsetlin (1962, 1966). It is characteristic of such a mode of control that the primary result of the "vertical" interaction of different levels of the hierarchical system is a change in the state principally (if not exclusively) of those "intermediate" (interneuronal) "computational" elements of the lower level that in turn provide for "horizontal" interactions of neuronal subsystems on this level. A change in the state of the output elements for this level is more the result of horizontal interaction than of direct vertical influences from higher levels on the lower-level output elements.

As our data indicate, the complex of spinal changes that precede the onset and determine the initiation of a voluntary movement may be divided into at least three independent processes, which we have termed "pretuning," "tuning," and "triggering" (Fig. 36). These processes differ in their time of appearance and in their duration. In all probability, they are determined by different mechanisms and are triggered by different supraspinal influences.

The "pretuning" process at the spinal level is manifested in an increase in the reflex excitability of the spinal motoneurons of the future agonist that occurs prior to presentation of the movement signal and remains unchanged for most of the latent period (see Chapter 3, Section 2).

The "pretuning" increase in the reflex excitability of the motoneuron pool of the future agonist is substantially greater than the increase in the reflex excitability of the other motoneuron pool that also occurs prior to the movement signal and remains constant during the entire (or most of the) latent period. In the latter case, the degree of the increase is the same for different motoneuron pools, regardless of the function of their muscles in the impending movement. Moreover, in a reaction involving a choice, in which case the function of a given muscle in the future movement is unknown prior to presentation of the signal, the early increase in the reflex excitability of the motoneuron pool of the muscle is roughly the same both when presentation of the signal makes it clear that the muscle is to act as agonist and when the muscle is not actively involved in the impending movement. The degree of the increase in reflex excitability in this case is approximately the same as in the motoneuron pools of the nonagonists in a reaction involving no choice.

These facts indicate that the increase in the reflex excitability of the motoneuron pools of muscles other than the agonists of the future movement as observed prior to presentation of the movement signal is a reflection of processes at the higher levels of the CNS that are associated with the state of attentiveness, signal expectancy, and a state of nonspecific readiness. In contrast to this "background," diffuse, "nonspecific" increase in the reflex excitability of different motoneuron pools, the "pretuning" increase in the reflex excitability of the motoneuron

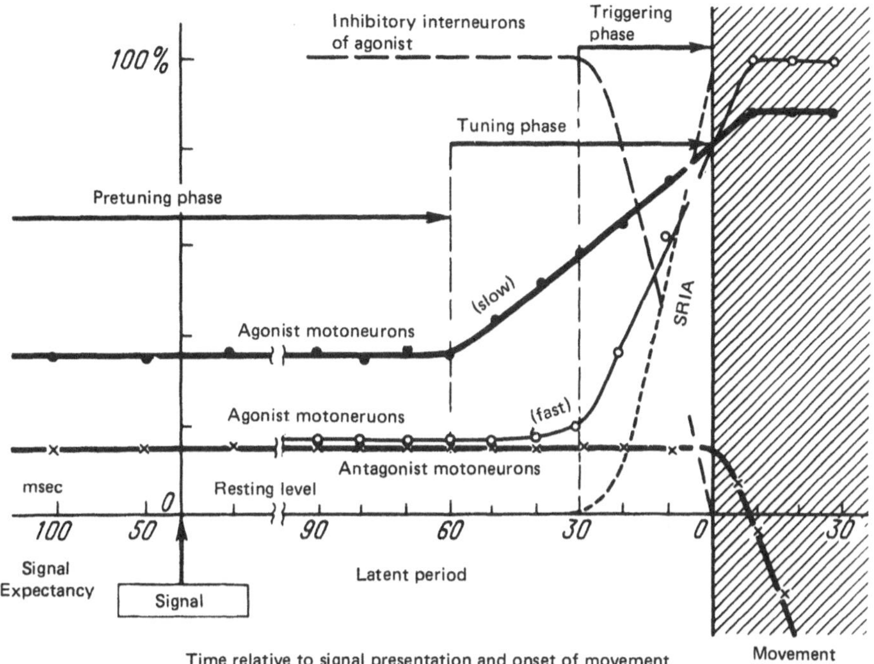

Fig. 36. Diagram of overall changes in the reflex excitability of the motoneuron pools of antagonist muscles and their associated spinal interneuronal apparatus during the organization of voluntary movement.

pools of the future agonists is characterized by such specific features that permit it to be associated with central (supraspinal) processes of the organization of a concrete voluntary movement. The latter conclusion is also substantiated by the fact that the "pretuning" increase in the reflex excitability of the agonist motoneuron pool is absent during the latent period of a reflex (involuntary) movement (see Chapter 5, Section 2).

Another clue to the specificity of "pretuning" may be seen in the fact that any spinal motoneuron pool of upper- or lower-limb musculature that is not reflexly activated on electrical stimulation of the UN during rest or during the performance of a voluntary movement not associated with the activity of this pool may be reflexly activated long in advance of a voluntary movement in which the motoneuron pool is to assume the role of primary agonist (see Chapter 5, Section 3).

It remains unclear which central structures and mechanisms are responsible for the "pretuning" increase in the reflex excitability of the motoneuron pools of future agonists. It is noteworthy, however, that the "pretuning" increase in the reflex excitability of the motoneuron pool of a paralyzed muscle is absent in patients with pyramidal paralysis of central origin during an attempt to perform a voluntary movement of the paralyzed limb, although a slight ("background") increase in the reflex excitability of the same motoneuron pool occurs during the preparation to perform a voluntary movement of the healthy leg (see Chapter 3, Section 2). This suggests that the systems and mechanisms responsible for the "pretuning" increase in the reflex excitability of spinal motoneurons are not identical to the systems and mechanisms that provide for the "background" (diffuse, "nonspecific") increase in their reflex excitability.

The "tuning" process at the spinal level is manifested in a smooth increase in the reflex excitability of the spinal motoneurons of the future agonist of a voluntary movement that precedes the onset of their impulse activity (see Chapter 3, Section 2). A characteristic feature of this supraspinal "tuning" of the reflex excitability of agonist motoneurons is its relatively long and constant duration: 55-60 msec. The duration of this supraspinal "tuning" is the same in subjects with different motor reaction times and is constant for repetitive movements (i.e., is the same during the performance of movements with different latencies). The duration of this process remains unchanged when the latency of a voluntary movement is shortened as a result of training, and it is the same during the performance of a phasic movement following rest (absence of prior impulse activity of motoneuron pool of future agonist of phasic movement) and during the organization of such a movement against a background of the voluntary tonic impulse activity of this motoneuron pool.

The constant duration of the "tuning" process appears either to be a reflection of the rigid temporal pattern of the supraspinal command or (which appears more likely) is associated with the constant time characteristics of the processes by which the supraspinal command is carried out at the segmental level. The second possibility is called into question by the fact that a smooth, 60-msec increase in the reflex excitability of the GM is observed not only prior to the onset of its voluntary impulse activity, but also in cases involving action from different reflex supraspinal influences--vestibulospinal and descending interlimb (see Chapter 5, Sections 1 and 2)--as well as in the presence of a loud audible stimulus (Zalkind and Shlykov, 1974).

Moreover, in cases in which electrical stimulation of the UN evokes the reflex impulse activity of the GM motoneurons ("ulnar" movement), the onset of this activity is also preceded by a "tuning" increase in the reflex excitability of the motoneuron pool of this muscle lasting about 60 msec, i.e., the same increase as recorded prior to voluntary movement, although the involuntary character of the "ulnar" movement is beyond question (see Chapter 5, Section 2).

Of course, since interlimb reflex influences are not observed in patients with pyramidal paralyses of voluntary movements that are central or spinal in origin, it cannot be ruled out that such reflex descending interlimb influences are, under normal conditions, associated at least partly with the reflex activation of the same central structures and are mediated by the same descending pathways that are also responsible for a voluntary command. If this is the case, the correspondence of the duration of the spinal "tuning" process in the case of voluntary and reflex "ulnar" movements may be determined by the common character of the supraspinal volley responsible for the origin of the spinal "tuning" preceding both movements. In this case, the hypothesis regarding the rigid temporal pattern of the supraspinal volley during a voluntary and a reflexly evoked movement seems less likely. In any case, such a hypothesis is contradicted by the sizable difference in the overall duration of the impulse activity of the GM motoneuron pool during these two movements.

As for vestibulospinal effects, they are preserved in patients with pyramidal paralyses of a central and spinal origin, and are absent after complete transection of the VIIIth nerve or are profoundly reduced in patients with lesions of the brain stem and the retention of voluntary movements (see Chapter 5, Section 1). For this reason, the similarity in the times of the "tuning" phase of the excitability increase in the GM motoneuron pool prior to voluntary movement or in response to stimulation of the UN, on the one hand, and in response to electrical stimulation of the vestibular apparatus, on the other, cannot be attributed to common central and descending supraspinal structures mediating all these effects. Moreover, it is hardly likely that the temporal pattern of the supraspinal command would be the same during the electrical stimulation of the vestibular apparatus with different strengths, polarities, and durations of the stimulus (from 1 to several dozen milliseconds), as it would be with the use of different sides of electrode attachment. In every case, the overall duration of the vestibulospinal facilitatory effect, the overall duration of the phase of the steadily increased motoneuronal excitability, and the duration of the phase of its decline change with changes in stimulus parameters. In all these cases, however, the duration of the first phase of the smooth ("tuning") excitability increase remains the same.

Thus, on the basis of the facts gathered, it may be supposed that the standard duration of the "tuning" increase in the reflex excitability of the agonist motoneuron pool prior to voluntary movement is determined by the temporal characteristics of the processes by which the supraspinal command is carried out at the segmental motor level, not by the rigid temporal characteristics of the supraspinal command itself.

The preliminary supraspinal activation of the spinal gamma loop has been suggested as one possible "candidate" for the mechanism responsible for the smooth "tuning" increase in the reflex excitability of the motoneuron pool of the future agonist. It has been demonstrated in experiments involving extremital ischemization, however, that a significant disruption in the afferent limb of the gamma loop has no appreciable effect on the character of the "tuning" excitability increase or the onset of the impulse activity of the motoneuron pool of the agonist of a voluntary movement (see Chapter 3, Section 3). Furthermore, if the "tuning" facilitation of agonist motoneurons were accomplished via the spinal gamma loop, one would expect a parallel inhibitory effect on the antagonist motoneuron pool due to reciprocal "direct" inhibition via the I-a afferents of the agonist. This is made more likely by the fact that the onset of voluntary movement is preceded by the supraspinal facilitation of the interneurons of "direct" antagonist inhibition (see Chapter 4, Section 3). However, as our findings show, no such inhibitory effect is observed during the last 60 msec of the latent period (see Chapter 4, Section 1). Our findings have led us to conclude that the peripheral gamma loop plays no critical role in the origin of the "tuning" reflex excitability increase or in the initiation of the voluntary impulse activity of the agonist motoneuron pool (see Chapter 3, Section 3).

We favor the hypothesis that both the "tuning" increase in the reflex excitability of the motoneuron pool of the agonist GM prior to voluntary movement and the initial phase of the smooth increase in the reflex excitability of this pool, owing to descending reflex influences, are associated with the activation of one and the same segmental intercalated (interneuronal) system, the time course and action of which on the GM motoneuron pool also determine the parameters of the "tuning." This system is activated either through different descending supraspinal (corticospinal, vestibulospinal, etc.) systems that converge at this segmental intercalated system, or through some single descending system. Since the latter is known to mediate vestibulospinal influences, it is entirely likely that the "tuning" increase in the reflex excitability of spinal motoneurons preceding the onset of voluntary movement is determined by the activity of a special segmental intercalated (interneuronal) system that is activated by the extrapyramidal (or "medial," after Kuypers) descending system.

Thus, it is our view that the "tuning" increase in the reflex excitability of the motoneuron pool of the agonist of a voluntary movement that precedes the onset of its impulse activity is associated with the extrapyramidal component of the voluntary (cortical) command. Both the extrapyramidal cortical system and the subcortical extrapyramidal system activated via the subcortical and stem collaterals of the corticospinal (pyramidal) tract may serve as the anatomical substrate of this component (Kuypers, 1958a,b, 1960, 1964; Patton and Amassian, 1960).

Such a hypothesis is supported in particular by the results of experiments by Ioffe (1970, 1973). Using the monosynaptic testing method, this investigator found that the onset of a conditioned reflex movement in dogs is also preceded by a "tuning" (our terminology) increase in the reflex excitability of the motoneurons (lasting approximately 80-100 msec). If a pyramidotomized animal is conditioned to perform such movements, their onset is also preceded by the usual phase of a "tuning" excitability increase, which suggests that the pyramidal tract need not be intact for the phenomenon of "tuning" to occur. It is interesting that in animals conditioned prior to pyramidotomy, the onset of voluntary movement is still preceded by the "tuning" increase in motoneuronal excitability, although in this case, the increase ceases 30-40 msec prior to movement.

Since a "tuning" increase in the reflex excitability of motoneurons is not observed in pyramidal patients with cortically localized lesions (see Chapter 3, Section 2), it may be concluded that under normal conditions, "tuning" supraspinal influences are determined by those cortical structures that are responsible for the central organization of voluntary movement, but are mediated by descending tracts belonging to the extrapyramidal, or medial (after Kuypers), system.

Our studies show that the "tuning" process, approximately 60 msec in duration, is apparently an obligatory spinal component of the central organization of any voluntary movement, and is not a phenomenon that is specific only with respect to the GM motoneuron pool. Indeed, the same character and duration are displayed by the "tuning" of the reflex excitability of the foot flexor motoneuron pool that precedes the onset of its voluntary impulse activity (see Chapter 4, Section 2), as well as of the motoneuron pool of the hand flexors (Mart'yanov and Kopylov, 1971). This again points to the universal character of the "tuning" mechanism, and the necessity of its participation in the organization of a phasic voluntary movement.

There is one final point that should be discussed with regard to the phenomenon of "tuning," namely, the unexpectedly high "inertness" of the spinal systems that carry out the supraspinal command. In fact, judging by the duration of "tuning," approximately 60 msec lapse from the first sign of the action of a supraspinal motor command on the segmental level to the onset of the realization of this command in the form of voluntary impulse activity of agonist motoneurons. It is noteworthy that this much time is consumed not only during the transition from a state of complete rest of the motoneuron pool to its impulse activity, but also when the motoneuron pool must change from one level of impulse activity to a different, higher level, as in the case of a phasic movement immediately following sustained muscular tension (see Chapter 3, Section 2).

Nothing we have learned to date from the results of experiments involving the electrical stimulation of various supraspinal motor structures in man and animals would have led us to expect such a lengthy spinal "delay" in the realization of supraspinal actions. It might have been suspected in all these cases, however, that the real times may have been different, it being unlikely that synchronous electrical stimulation would produce an accurate picture of the activation of these structures. In fact, when the activity of the pyramidal tract neurons--the output neurons of the motor cortex in monkeys--was recorded during the performance of natural movements, it was found that a change in their activity precedes the onset of the impulse activity of the spinal motoneurons by several dozen milliseconds (see Chapter 1, Section 1).

The presence of the phenomenon of a prolonged spinal delay makes it possible to examine one important aspect of the temporal characteristics of the supraspinal control of movement. It may be assumed that a simple, rapid, single voluntary movement in one joint with no special requirements in terms of kinematic or dynamic parameters (accuracy of amplitude, speed of movement, muscular effort) is the result of the realization of a "unitary" supraspinal (cortical) command (Wagner, 1925; Kosilov and Farfel', 1938; Alekseev and Asknazii, 1970). At the same time, it is discovered that the realization of this "unitary" command includes a 60-msec delay at the spinal level. If a spinal delay of this duration is necessary for the realization of any "unitary" supraspinal command that is formulated in the process of motor control (Bernstein, 1935, 1947, 1966), it follows that each command can be carried out by the spinal cord only after a relatively long period of time. It is therefore quite likely that the temporal characteristics of the segmental level of motor control limit the frequency of "intervention" of the supraspinal motor centers in the work of this lower level, thereby giving the supraspinal control an intermittent character.

The hypothesis regarding the discontinuous ("quantum") control of movements has emerged many times in the literature in one form or another (Bernstein, 1966; Aizerman and Andreeva, 1970; Gidikov, 1970; and others). For example, while discussing the question of the time of the correction ("flowing microregulation") of a movement, Bernstein (1966) suggested the possible "quantization" of sensory and coordinative processes at time intervals on the order of 0.07-0.12 sec. According to Bernstein's hypothesis, this quantization frequency may reflect rhythmic fluctuations in the excitability of "all, or at least the principal, elements of the reflex ring of the control system of the human motor apparatus" (p. 86). There is a considerable volume of experimental neurophysiological data indicating that such quantization may actually take place in various limbs of the motor control system (Vasilenko and Kostyuk, 1966; Sverdlov, S. M., and Maksimova, 1966; Sweet and Bourassa, 1967).

The "triggering" process at the spinal level covers approximately the last 30 msec of the latent period preceding the onset of a voluntary movement. This time interval is characterized by a rapid increase in the reflex excitability of the "fast" motoneurons of the agonist GM. Experiments with animals, as well as our findings (see Chapter 2, Section 4, and Chapter 3, Section 2), suggest that this phenomenon may be the result of pyramidal "triggering" influences. The high effectiveness of pyramidal facilitatory actions on the "fast" motoneurons and the low effectiveness on the "slow" motoneurons of the GM satisfactorily explain why the sharp "triggering" increase in the reflex excitability of the "fast" motoneurons during the last 30 msec of the latent period is not accompanied by a parallel increase in the rate of rise of the reflex excitability of the "slow" GM motoneurons.

One component of the spinal changes occurring during the organization of voluntary movement is a depression of the inhibitory interneuronal system acting on the motoneuron pool of the future agonist (see Chapter 3, Section 4). Since this depression takes place during the last 25-30 msec of the latent period, it can presumably be linked to the same pyramidal "triggering" influences that determine the sharp increase in the reflex excitability of "fast" motoneurons that also occurs during this interval. The hypothesis regarding the essential role of "triggering" pyramidal influences in the changes in the segmental interneuronal apparatus is consistent with experimental data showing that the pyramidal system has effective ties with the interneuronal apparatus of the spinal cord (see Chapter 3, Section 4, and Chapter 4, Section 3).

Thus, during the last period of the organization of a voluntary phasic movement, the motoneuron pool of the future agonist of the movement is apparently the recipient of at least two types of influence: "tuning" and "triggering." The first of these influences determines the "tuning" increase in the reflex excitability of the agonist motoneuron pool that begins 55-60 msec prior to the onset of its impulse activity, and is probably determined by the extrapyramidal component of the supraspinal command. The "triggering" influences are reflected in the sharp selective increase in the reflex excitability of the "fast" motoneurons of the agonist motoneuron pool that begins 25-30 msec prior to its impulse activity. It is likely that these latter influences are mediated by the pyramidal system.

Such a conclusion appears to be in agreement with a widely accepted hypothesis regarding the participation of two systems--the pyramidal and extrapyramidal--in the control of movement, most physiologists agreeing that both the pyramidal and extrapyramidal systems are involved in the control of any voluntary movement. However, the roles of these two systems in this control are clearly different. The extrapyramidal system controls essentially the postural activity of axial and proximal limb musculature, ensuring a "background" against which the pyramidal system controls discrete voluntary movements executed by distal limb musculature.

Critically analyzing the tremendous volume of experimental and clinical material that had accumulated by the end of the 1930's, N. A. Bernstein wrote: "Long ago, when the ideas of Gitzig and Munk were predominant and localizationism was the fashion, a movement was regarded as a one-voiced melody performed by the muscles under the direction of the cellular 'keyboard' of the cerebral cortex. A movement was considered a relatively simple event: for a joint to be bent, an impulse must be sent to the proper flexor.

"As the concept of motor function grew more complex, and the notion of a biomechanical hierarchy was understood, the structural complexity of its anatomical foundation grew increasingly apparent.

"At present, it is generally agreed that the extrapyramidal system is of profound significance, even in man. It is true that with the emergence and development of the newer, pyramidal motor system, the primitive extrapyramidal system was 'shoved' aside and relegated to the role of an indistinct background against which the more salient motor manifestations of the pyramidal apparatus occur. However, every voluntary movement requires a background or a 'context' in order to exist. Every complete motor act is innervated by both systems, and the basic elements of the activity are inherent in each. The extrapyramidal system is, in fact, the predominant participant in many movements, with the cerebral cortex ready at any moment to add a single purposive act" (Bernstein, 1936, manuscript of the State Central Institute of Physical Culture).

The most characteristic features of the extrapyramidal control of movement are the following: (1) its actions are directed predominantly (if not exclusively) toward axial and proximal limb musculature; (2) it simultaneously activates (or inhibits) large, synergistic muscle groups; (3) the activity of the muscles controlled by extrapyramidal mechanisms is "tonic" in character; (4) it provides for postural regulation, but not for phasic movement.

For example, Hines (1943) turned her attention to the findings of Tower (1940), according to which integrated "synergistic movements" involving primarily axial and proximal limb musculature are evoked in the pyramidotomized monkey by the stimulation of cortical motor field 4. Extrapyramidal movements such as these are characterized by a slow onset and delayed cessation. Hines called this type of generalized muscular activity controlled by extrapyramidal mechanisms "holokinesis," thereby emphasizing its similarity to the movements of protozoans and higher animals at the early stages of ontogenesis. According to Hines's hypothesis, these stereotype movements that are integrated at the lower levels are utilized as parts of postural adaptations and thereby ensure the realization of pyramidal control. "Idiokinesis" was the term Hines gave to the precision that the pyramidal system adds to holokinesis.

The theory of Hess (1949, 1954) is very much in line with the concept of Hines. According to Hess, a specific starting position must be assumed in order to perform a purposive movement. The reflex preparations of a starting position ensure a "dynamic substrate" onto which voluntary purposive movements are impressed. "Teleokinetic" was the term Hess gave to the voluntarily controlled phase of movement, while the other phase, which ensures the underlying conditions for each precise purposive movement, was called the "ereismatic" phase.

Building on Hess's theory, Jang and Hassler (1960) point out that ereismatic regulation not only precedes a movement, but also acts later to compensate for unforeseen recoil reactions stemming from reactive forces. As an example, Jang and Hassler present the following observations: An intentional (voluntary) body turn in man determines this sequence of events: an eye movement, followed by a head turn and then by a body movement; passive repetitive movements are accompanied by these movements in reverse order: first the body moves, followed by the head and then by the eyes.

Jung and Hassler interpret this fact as follows: The performance of a voluntary movement that is primarily teleokinetic is accompanied by the physiological anticipation of the motor activity, including its sensory component (eye movement as leading sensory input for future movement)--sensory anticipation--and the activation of protective mechanisms, which are components of motor anticipation. Vestibular corrective movements evoked by passive angular acceleration appear to be a manifestation of those ereismatic mechanisms that are directed toward the maintenance of a normal posture. Jang and Hassler suggest that the functions of motor readiness, anticipation, or attention must work in close connection with the nonspecific activating system of the brain stem.

Judging by the results of apraxia brought on by lesions of the cerebral cortex, leading to the impairment of this anticipatory activity, Jang and Hassler hypothesize that the cerebral cortex and the thalamocortical connections play a leading role in motor anticipation. The extrapyramidal centers, reticular formation, and spinal gamma loop appear to be the effector mechanisms of this function.

In a comprehensive survey of previous studies, Paillard (1960), with reference to Hess, points out that the extrapyramidal system provides an essential framework for economical and precise execution of purposive ("teleokinetic") movements. In the interest of accuracy, it should be noted that Hess himself emphasized that his functional conception includes no differentiation between pyramidal and extrapyramidal innervation, since these terms are intended to describe anatomical structures rather than the two forms of motor regulation the existence of which is postulated in his hypothesis.

In a survey published in 1967, Buchwald (1967) defines the "classic" concept of motor control according to this concept: the pyramidal system is considered responsible for the ability to perform voluntary movements by the activation of particular muscles and the inhibition or graded relaxation of their antagonists, and by the combination of contractions in many different muscle groups and the inhibition of their antagonists. The extrapyramidal system is believed responsible for the activation of large muscle groups and the depression of the activity of other groups, thereby generating stereotype motor patterns.

There has been a tendency in recent years to depart from the traditional division of the central motor control system into the pyramidal and extrapyramidal subsystems in favor of a new recombination of the basic descending motor systems into two other subsystems: the lateral and medial (Kuypers, 1964; Lawrence and Kuypers, 1968a,b). Nevertheless, the same functions are ascribed to these two systems as were previously associated with the pyramidal and extrapyramidal systems, respectively. For example, Kuypers (1964) states that the medial system controls basically axial and proximal limb musculature (particularly extensors) and fulfills postural and locomotive functions, while the lateral system is particularly important in the control of distal limb musculature (notably the flexors) and provides for discrete coordinated movements.

We believe, however, that results obtained in the present investigation conflict in several respects with classic notions regarding the character of the participation and functional role of the pyramidal and extrapyramidal systems in the organization of voluntary movement. First, the extrapyramidal (according to our hypothesis) "tuning" influences that precede the onset of voluntary movement are strictly localized: they are limited to the motoneuron pools (or the associated interneuronal nuclei) of the primary agonists of the future movement. The classic theory, on the other hand, postulates the participation of the extrapyramidal system in the regulation of the activity of muscles other than the primary agonists, notably muscles providing for postural fixation and compensation for postural disruptions during the execution of a movement.

Furthermore, the classic theory holds that the extrapyramidal system essentially controls axial musculature (and proximal limb muscles), particularly the extensors. According to our data, the phenomenon of the (extrapyramidal) "tuning" is observed in the agonist motoneuron pool of the flexor (ATM) as well as in the extensor (GM). It is noteworthy that both these muscle groups belong to the distal musculature of the lower limb. Moreover, the phenomenon of "tuning" is also observed prior to the onset of the voluntary impulse activity of the motoneuron pools of hand and forearm muscles (Zalkind, unpublished communication; Mart'yanov and Kopylov, 1971), muscles that, according to the classic theory, lie largely outside the sphere of control of the extrapyramidal system.

The data obtained cast no doubt on the possible role of the extrapyramidal system in providing a postural "background" against which phasic voluntary purposive movements controlled primarily by the pyramidal system are performed. It is quite probable that the extrapyramidal system also provides for the postural readjustments that precede the onset of a voluntary movement (Hess, 1949, 1954; Kas'yanov, 1950; Jang and Hassler, 1960; Belen'kii et al., 1967). Supplementing these classic conceptions of the spatial (different muscle groups) and temporal (postural readjustments followed by movement) separation of the functions of the extrapyramidal and pyramidal systems, our findings indicate that the organization of the activity of the primary agonists of a voluntary movement is apparently associated with the participation of at least two descending types of influence on the segmental motoneuron pools of these muscles. The hypothesis that these two types of influences are mediated by the extrapyramidal and pyramidal systems is quite probable. The participation of both systems in the initiation of voluntary movement makes it clear why "pure" damage to one (in particular the pyramidal pathway) does not destroy the ability to perform voluntary movements in primates (Tower, 1940; Liu and Chambers, 1964; Denny-Brown, 1966; Bucy et al., 1966), including man (Lassek, 1954; Bucy et al., 1964).

The existence of a final common pathway for pyramidal and extrapyramidal actions on the motoneuron pool of the agonists of a voluntary movement raises important questions concerning the character of the interaction of these two types of descending influence during the organization and realization of a voluntary movement. We take note in this regard of the results of our experiments that were designed to investigate the effect of descending reflex vestibulospinal (extrapyramidal) influences during the organization of voluntary movement (see Chapter 5, Section 3). We found that during the last 30 msec of the latent period and the first 30 msec of the voluntary impulse activity of the agonist motoneuron pool, there is a complete "switch-off" (blockade) of direct vestibulospinal influences on this motoneuron pool.

On the basis of this fact, it may be supposed that during the appearance of "triggering" (pyramidal) influences, there is no chance of direct actions on the motoneuron pool of the active muscles via descending vestibulospinal and other extrapyramidal (Zalkind, 1972) as well as segmental (Korobkov et al., 1969) reflex pathways.

Thus, during this period, the control of the motoneuron pools of the primary agonists of a voluntary movement evidently becomes the exclusive function of the corticospinal system. However, there is not yet enough factual evidence on which to base adequately specific judgments on the character of the interaction of the pyramidal and extrapyramidal systems during the organization and execution of a voluntary movement.

It should be emphasized that the characteristics of this interaction that were observed in our experiments may have relevance only to the simplest class of voluntary movement, which was the object of the present study. On the other hand, it is possible that the transient "switch-off" (blockade) of vestibulospinal facilitatory actions on the motoneuron pool of the agonist of a voluntary movement is a partial manifestation of a general mechanism for the temporary "escape" of the active segmental level from any (also pyramidal) supraspinal influences during the realization of a "unitary" supraspinal command for a voluntary movement. In this case, the 60-msec period for which the vestibulospinal facilitatory effect is absent reflects the duration of the spinal realization of the supraspinal command (duration of the spinal "cycle" or a "quantum" of motor control).

Even as one "unitary" supraspinal command is being realized, the next "unitary" supraspinal command can evoke new changes in the segmental apparatus that, after some delay, will be realized by this apparatus as a change in its output (motoneuronal activity). The spinal organization of a new voluntary movement (or a part of it) may occur during the performance of the preceding movement. In fact, during the performance of two successive movements (see Chapter 4, Section 1), a change in the reflex excitability of the motoneuron pool of the agonist of the second movement begins during the performance of the first movement, and the onset of these changes is independent of the presence or absence of the impulse activity of antagonist motoneuron pools during this period.

Thus, the smoothness of the execution of central movements, regardless of the cyclic ("quantized") character of their central control, may be due to the fact that the next part of a movement is already being organized during the execution of the preceding part.

REFERENCES

Abrahams, V. C., "Cervico-lumbar reflex interactions involving a proprioceptive receiving area of the cerebral cortex," J. Physiol. (England) 209:45-56 (1970).

Adamovich, N. A., and Borgest, A. N., "On the interaction of segmental and intersegmental cord reflexes," Fiziol. Zh. SSSR im. I. M. Sechenova 54:787-795 (1968).

Adamovich, N. A., Borgest, A. N., and Evdokimov, S. A., "The effects of afferent impulses of forelimb nerves on interneurons of lumbar region of spinal cord," Neurofiziologiya 1:235-242 (1969a).

Adamovich, N. A., Borgest, A. N., and Evdokimov, S. A., "The influence of forelimb afferents on the 'spontaneous' activity of interneurons of lumbar cord," Dokl. Akad. Nauk SSSR 184:493-496 (1969b).

Agnew, R. F., and Preston, J. B., "Motor cortex-pyramidal effects on single ankle flexor and extensor motoneurons of the cat," Exp. Neurol. 12:384-398 (1965).

Agnew, R. F., Preston, J. B., and Whitlock, D. G., "Patterns of motor cortex effects on ankle flexor and extensor motoneurons in the 'pyramidal' cat preparation," Exp. Neurol. 8:248-263 (1963).

Ainsworth, A., Gaffan, G. D., O'Keefe, J., and Sampson, R., "A technique for recording units in the medulla of the awake freely moving rat," J. Physiol. (England) 202:80-82 (1969).

Aizerman, M. A., and Andreeva, E. A., On Some Mechanisms of the Control of Skeletal Muscles [in Russian], Moscow, Science Publishing House, 1968.

Aizerman, M. A., and Andreeva, E. A., "On some of the simplest mechanisms of the control of skeletal muscles," in: Investigation of the Processes of the Control of Muscular Activity [in Russian], Moscow, Science Publishing House, 1970, pp. 5-49.

Albe-Fessard, D., "Organization of somatic central projections," in: Contributions to Sensory Physiology, Vol. 2, New York, Academic Press, 1967, pp. 101-169.

Alekseev, M. A., "Systemic activity of the higher areas of the brain and some questions on the subject of human motor control," Zh. Vyssh. Nervn. Deyat. im. I. P. Pavlova 17:786-797 (1967).

Alekseev, M. A., and Asknazii, A. A., "Some principles of the control of precise cyclic movements in man," in: The Control of Movements [in Russian], Leningrad, Science Publishing House, 1970, pp. 17-37.

Amatuni, A. S., "The influence of the anterior lobe of the cerebellum on monosynaptic reflex cord reactions in the precollicular cat," Zh. Vyssh. Nervn. Deyat. im. I. P. Pavlova 17:1098-1105 (1967).

Andersen, P., Eccles, J. C., and Sears, T. A., "Cortically evoked depolarization of primary afferent fibers in the spinal cord," J. Neurophysiol. 27:63-77 (1964a).

Andersen, P., Eccles, J. C., Oshima, T., and Schmidt, R. F., "Mechanisms of synaptic transmission in the cuneate nucleus," J. Neurophysiol. 27:1096-1116 (1964b).

Andersson, S., and Gernandt, B. E., "Ventral root discharge in response to vestibular and proprioceptive stimulation," J. Neurophysiol. 19:524-543 (1956).

Angel, R. W., Garland, H., and Alston, W., "Interaction of spinal and supraspinal mechanisms during voluntary innervation of human muscle," Exp. Neurol. 28:230-242 (1970).

Appelberg, B., "The effect of electrical stimulation of nucleus ruber on the gamma motor system," Acta Physiol. Scand. 55:150-159 (1962).

Appelberg, B., and Emonet-Denand, F., "Central control of static and dynamic sensitivities of muscle spindle primary endings," Acta Physiol. Scand. 63:487-494 (1965).

Appelberg, B., and Molander, C., "A rubro-olivary pathway. I. Identification of a descending system for control of the dynamic sensitivity of muscle spindles," Exp. Brain Res. 3:372-381 (1967).

Asanuma, H., and Rosen, I., "Topographical organization of cervical efferent zone projections to distal forelimb muscles in the monkey," Exp. Brain Res. 14:243-256 (1972).

Asanuma, H., and Sakata, H., "Functional organization of a cortical efferent system examined with focal depth stimulation in cats," J. Neurophysiol. 30:35-54 (1967).

Asanuma, H., and Ward, J. E., "Patterns of contraction of distal forelimb muscles produced by intracortical stimulation in cats," Brain Res. 27:97-109 (1971).

Asratyan, E. A., Physiology of the Central Nervous System [in Russian], Moscow, Academy of Medical Sciences of the USSR, 1953.

Avakyan, R. V., Vardanetyan, G. A., and Gershuni, G. V., "Investigation of the latent period of a voluntary motor reaction in response to audible signals of varying duration and intensity," Zh. Vyssh. Nervn. Deyat. im. I. P. Pavlova 16:1037-1045 (1966).

Babkin, P. S., "Physiological variants of proprioceptive and exteroceptive reflexes in man" (Doctoral dissertation), Moscow (1967).

Baikushev, St., Manovich, Z. Kh., and Novikova, V. P., Stimulative Electromyography and Electroneurography in the Diagnosis and Treatment of Nervous Diseases [in Russian], Moscow, Medicine Publishing House, 1974.

Baranov-Krylov, I. N., "On the excitability testing of the spinal centers for antagonist muscle pairs," Zh. Vyssh. Nervn. Deyat. im. I. M. Pavlova 19:889-891 (1969).

Barnes, C. D., and Pompeiano, O., "Presynaptic inhibition of extensor monosynaptic reflex by la afferents from flexors," Brain Res. 18:380-383 (1970).

Bassin, F. V., and Serkova, M. P., "On the electrography of changes in muscle tonus preceding a voluntary movement in the presence of organic disorders of the motor system," Zh. Nevropatol. Psikhiatr. im. S. S. Korsakova 56:866-873 (1956).

Bassin, F. V., and Sidorov, P. I., "Electromyographic analysis of the readiness to move in normal subjects, and in the presence of cerebrovascular pathology and other afflictions of the CNS," in: IV-th All-Union Congress of Neuropathologists and Psychiatrists [in Russian] Vol. 1, Moscow, State Medical Publishing House, 1963, p. 267.

Batini, C., Moruzzi, G., and Pompeiano, O., "Cerebellar release phenomena," Arch. Ital. Biol. 95:71-95 (1957).

Batuev, A. S., Functions of the Motor Analyzer [in Russian], Leningrad State University Press, 1970.

Bava, A., Manzoni, T., and Urbano, A., "Cerebellar influences on neuronal elements of thalamic somatosensory relay-nuclei," Arch. Sci. Biol. 50:181-204 (1966).

Bekhterev, V. M., The Fundamentals of Brain Functions [in Russian], 4th Ed., St. Petersburg, 1905.

Bekhtereva, N. P., Neurophysiological Aspects of Psychic Activity in Man [in Russian], Leningrad, Medicine Publishing House, 1971.

Bekhtereva, N. P., Bondarchuk, A. N., Smirnov, V. M., and Trokhachev, A. I., The Physiology and Pathopsysiology of Deep Brain Structures in Man [in Russian], Leningrad, Moscow, Medicine Publishing House, 1967.

Belekhova, M. G., "Post-tetanic potentiation," Usp. Sovrem. Biol. 66(2):199-225 (1968).

Belen'kii, V. E., Gurfinkel', V. S., and Pal'tsev, E. I., "On the elements of voluntary motor control," Biofizika 12:135-141 (1967).

Bergmans, J., and Colle, J., "Etude des phênomènes d'inhibition présynaptique au niveau des régions cervicale et lambaire de la moelle épinière chez la grenouille," Arch. Int. Physiol. 72:724-726 (1964).

Beritov, I. S., Study on the Basic Elements of the Central Coordination of Skeletal Musculature [in Russian], Petersburg, 1916.

Bernhard, C. G., and Bohm, E., "Cortical representation and functional significance of the cortico-motoneuronal system," Arch. Neurol. Psychiatry 72:473-502 (1954).

Bernhard, C. G., Bohm, E., and Petersen, I., "Investigations on the organization of the cortico-spinal system in monkeys (*Macaca mulata*)," Acta Physiol. Scand. 29 (Suppl. 106):79-103 (1953).

Bernstein [Bernshtein], N. A., "The problem of the interrelations of coordination and localization," Arkh. Biol. Nauk 38(1):1-38 (1935).

Bernstein [Bernshtein], N. A., On the Structure of Movements, [in Russian], Moscow, State Medical Publishing House, 1947.

Bernstein [Bernshtein], N. A., "Methods and problems in the physiology of activity," Vopr. Filos., No. 6, 77-92 (1961).

Bernstein [Bernshtein], N. A., Essays on the Physiology of Movement and the Physiology of Activity [in Russian], Moscow, Medicine Publishing House, 1966.

Blinkov, S. M., and Ponomarev, V., "Quantitative determination of neurons and glial cells in the nuclei of the facial and vestibular nerves in man, monkey and dog," J. Comp. Neurol. 125:295-301 (1965).

Blinkov, S. M., Arutyunova, A. S., and Moskatova, A. K., "Reaction time and brain structure," in: Psychological Investigations [in Russian], No. 1, Moscow State University Press, 1968, pp. 60-71.

Blom, S., Hagbarth, K. E., and Skoglund, S., "Post-tetanic potentiation of H-reflexes in human infants," Exp. Neurol. 9:198-211 (1964).

Boiko, E. I., Human Reaction Time [in Russian], Moscow, Medicine Publishing House, 1964.

Brodal, A., The Reticular Formation of the Brain Stem, Springfield, Illinois, Charles C. Thomas, 1957.

Brodal, A., Val'berg, F., and Pompeiano, O., Vestibular Nuclei and Their Connections, Anatomy and Functional Correlations, Springfield, Illinois, Charles C. Thomas, 1962.

Brooks, C. M., Downman, C. B. B., and Eccles, J. C., "After-potentials and excitability of spinal motoneurons following orthodromic activation," J. Neurophysiol. 13:159-176 (1950).

Brooks, V. B., and Stoney, S. D., "Motor mechanisms: The role of the pyramidal system in motor control," Annu. Rev. Physiol. 33:337-392 (1971).

Brooks, V. B., Adrien, J., and Dykes, R. W., "Task-related discharge of neurons in motor cortex and effects of dentate cooling," Brain Res. 40:85-88 (1972).

Bruggencate, G., Burke, R., and Lundberg, A., "Interaction between the vestibulospinal tract, contralateral flexor reflex afferents and Ia afferents," Brain Res. 14:529-532 (1969).

Brunia, C. H. M., "The influence of a task on the Achilles tendon and Hoffmann reflex," Physiol. Behav. 6:367-373 (1971).

Buchthal, F., Guld, C., and Rosenfalck, P., "Propagation velocity in electrically activated muscle fibers in man," Acta Physiol. Scand. 34:75-89 (1955).

Buchwald, J. S., "A functional concept of motor control," Am. J. Phys. Med. 46:141-150 (1967).

Buchwald, J. S., Beatty, D., and Eldred, E., "Conditioned responses of gamma- and alpha-motoneurons in the cat trained to conditioned avoidance," Exp. Neurol. 4:91-109 (1961).

Buchwald, J. S., Standish, M., and Eldred, E., "Effect of deafferentation upon acquisition of a conditioned flexion response in the cat," Exp. Neurol. 9:372-385 (1964).

Bucy, P. C., Keplinger, J. E., and Siqueira, E. B., "Destruction of the 'pyramidal tract' in man," J. Neurosurg. 21:385-398 (1964).

Bucy, P. C., Ladpli, R., and Ehrlich, A., "Destruction of the pyramidal tract in the monkey," J. Neurosurg. 25:1-23 (1966).

Burg, D., Szumski, A. J., Struppler, A., and Velho, F., "Afferent and efferent activation of human muscle receptors involved in reflex and voluntary contraction," Exp. Neurol. 41:754-768 (1973).

Burke, R. E., "Group Ia synaptic input to fast and slow twitch motor units of cat triceps surae," J. Physiol. (England) 196:605-630 (1968).

Buser, P., Ascher, P., Bruner, J., Jassik-Gerschenfeld, D., and Sindberg, R., "Aspects of sensorimotor reverberation to acoustic and visual stimuli," in: Brain Mechanisms (G. Moruzzi et al., eds.), Amsterdam-London-New York, Elsevier, 1963, pp. 294-324.

Byuzer, P., and Ember, M., "Sensory projections in the cat motor cortex," in: The Theory of Connection in Sensory Systems [in Russian], Moscow, World Publishing House, 1964, pp. 214-231.

Calma, I., and Kidd, G. L., "The action of the anterior lobe of the cerebellum on alpha-motoneurons," J. Physiol. (England) 149:626-652 (1959).

Carli, G., Diete-Spiff, K., and Pompeiano, O., "Presynaptic and postsynaptic inhibition of transmission of somatic afferent volleys through the cuneate nucleus during sleep," Arch. Ital. Biol. 105:52-82 (1967).

Carpenter, D., Lundberg, A., and Norsell, U., "Effects from the pyramidal tract on primary afferents and on spinal reflex actions to primary afferents," Experientia 18:337-338 (1962a).

Carpenter, D., Engberg, I., and Lundberg, A., "Presynaptic inhibition in the lumbar cord evoked from the brain stem," Experientia 18:450-451 (1962b).

Carpenter, D., Lundberg, A., and Norsell, U., "Primary afferent depolarization evoked from the sensorimotor cortex," Acta Physiol. Scand. 59:126-142 (1963).

Carpenter, D., Engberg, I., and Lundberg, A., "Primary afferent depolarization evoked from the brain stem and the cerebellum," Arch. Ital. Biol. 104:73-85 (1966).

Chambers, W. W., and Liu, C. N., "Cortico-spinal tract in monkey," Fed. Proc. Fed. Am. Soc. Exp. Biol. 17:24-51 (1958).

Cherkes, V. A., Essays on the Physiology of the Basal Ganglia of the Brain [in Russian], Kiev, 1963.

Chuprikova, N. I., Speech As a Control Factor in Higher Nervous Activity in Man [in Russian], Moscow, Education Publishing House, 1967.

Close, R. I., "Dynamic properties of mammalian skeletal muscles," Physiol. Rev. 52:129-197 (1972).

Clough, J. F. H., Kernell, D., and Phillips, C. G., "The distribution of monosynaptic excitation from the pyramidal and from primary spindle afferents to motoneurons of the baboon hand and forearm," J. Physiol. (England) 198:145-166 (1968).

Cook, W. A., Cangiano, A., and Pompeiano, O., "Vestibular influences on primary afferents in the spinal cord," Pfluegers Arch. Gesamte Physiol. Menschen Tiere 299:334-338 (1968).

Coquery, J. M., "Changes in somaesthetic evoked potentials during movement," Brain Res. 31:375 (1971).

Coquery, J. M., and Coulmance, M., "Variations d'amplitude des réflexes monosynaptiques avant un mouvement volantaire," Physiol. Behav. 6(1):65-69 (1971).

Corazza, R., Fadiga, E., and Parmeggiani, P. L., "Patterns of pyramidal activation of cat's motoneurons," Arch. Ital. Biol. 10:337-364 (1963).

Corrie, W. S., and Hardin, W. B., "Post-tetanic potentiation of H-reflex in normal man," Arch. Neurol. 11:317-323 (1964).

Coulter, J. D., and Thies, R., "Sensory transmission through the lemniscal pathway during movements in arousal in the cat," Fed. Proc. Fed. Am. Soc. Exp. Biol. 30:664 (1971).

Creed, R. S., Denny-Brown, D., Eccles J., Luddell, E., and Sherrington, C., Reflex Activity of the Spinal Cord, Fair Lawn, New Jersey, Oxford University Press, 1932.

Deecke, L., Scheid, P., and Kornhuber, H. H., "Distribution of readiness potential, pre-motion positivity, and motor potential of the human cerebral cortex preceding voluntary finger movements," Exp. Brain Res. 7:158-168 (1969).

De Long, M. R., "Activity of basal ganglia neurons prior to movement," Fed. Proc. Fed. Am. Soc. Exp. Biol. 30:433 (1971a).

De Long, M. R., "Activity of pallidal neurons during movement," J. Neurophysiol. 34:414-427 (1971b).

De Long, M. R., "Activity of basal ganglia neurons during movement," Brain Res. 40:127-135 (1972).

Denny-Brown, D., "On the nature of postural reflexes," Proc. R. Soc. London 104:252-301 (1929).

Denny-Brown, D., The Basal Ganglia, London, Oxford University Press, 1962.

Denny-Brown, D., The Cerebral Control of Movement, Liverpool University Press, 1966.

Diamantopoulos, E., and Gassel, M. M., "Electrically induced monosynaptic reflexes in man," J. Neurol. Neurosurg. Psychiatry 28: 496-502 (1965).

Diete-Spiff, K., Carli, G., and Pompeiano, O., "Spindle responses and extrafusal contraction on stimulation of the VIIIth cranial nerve or the vestibular nuclei in the cat," Pfluegers Arch. Gesamte Physiol. Menschen Tiere 293:276-280 (1967).

Dobronravova, I. S., "Electroencephalographic characteristics of the different stages of a conditioned motor reaction in man," in: The Neural Mechanisms of Conditioned Reflex Activity [in Russian], Moscow, Science Publishing House, 1963, pp. 118-126.

Dow, R. S., and Moruzzi, G., The Physiology and Pathology of the Cerebellum, Minneapolis, University of Minnesota Press, 1958.

Drozdova, V. N., "The electromyogram of the dorsal roots of dogs in intact and completely deafferented preparations," in: Mechanisms of Compensative Adaptations [in Russian], Moscow, Science Publishing House, 1964, pp. 104-112.

Drozdova, V. N., "On the analysis of the electrical muscular activity of deafferented and intact dog dorsal roots on stimulation of the motor area of the cerebral cortex," in: The Neural Mechanisms of Motor Activity [in Russian], Moscow, Science Publishing House, 1966, pp. 153-158.

Eccles, J. C., Physiology of Nerve Cells, Baltimore, Johns Hopkins Press, 1957.

Eccles, J., Physiology of Synapses, New York, Academic Press, 1964.

Eccles, J., Inhibitory Pathways of the Central Nervous System, Springfield, Illinois, Charles C. Thomas, 1969.

Eccles, J. C., Fatt, P., and Koketsu, K., "Cholinergic and inhibitory synapses in a pathway from motor-axon collaterals to motoneurons," J. Physiol. (England) 126:524-562 (1954).

Eccles, J. C., Eccles, R. M., and Lundberg, A., "The convergence of monosynaptic excitatory afferents on to the many different species of alpha-motoneurons," J. Physiol. (England) 137:22-50 (1957).

Eccles, J. C., Eccles, R. M., and Lundberg, A., "The action potentials of the alpha motoneurons supplying fast and slow muscles," J. Physiol. (England) 142:275-291 (1958).

Eccles, J. C., Eccles, R. M., Iggo, A., and Ito, M., "Distribution of recurrent inhibition among motoneurons," J. Physiol. (England) 159:479-499 (1961).

Eccles, J. C., Kostyuk, P. G., and Schmidt, R. F., "Presysnaptic inhibition of the central actions of flexor reflex afferents," J. Physiol. (England) 161:258-281 (1962).

Eccles, J. C., Ito, M., and Szenthagothai, J., The Cerebellum As a Neuronal Machine, Berlin-Heidelberg-New York, Springer-Verlag, 1967.

Eccles, J. C., Saban, M. H., Schmidt, R. G., and Taborikova, H., "Mode of operation of the cerebellum in the dynamic loop control of movement," Brain Res. 40:73-80 (1972).

Eccles, R. M., and Lundberg, A., "Supraspinal control of interneurons mediating spinal reflexes," J. Physiol. (England) 147:565-584 (1959a).

Eccles, R. M., and Lundberg, A., "Synaptic actions in motoneurons by afferents which may evoke the flexion reflex," Arch. Ital. Biol. 97:199-221 (1959b).

Ekholm, J., and Skoglund, S., "Possible factors influencing the demonstration of post-tetanic potentiation of the H-reflex as studied in the cat," Exp. Neurol. 9:183-197 (1964).

Eldred, E., Schnitzlein, H. N., and Buchwald, J., "Response of muscle spindles to stimulation of the sympathetic trunk," Exp. Neurol. 2:13-25 (1960).

Engberg, I., Lundberg, A., and Ryall, R. W., "Reticulo-spinal inhibition of transmission through interneurons of spinal reflex pathways," Experientia 21:612-613 (1965).

Engberg, I., Lundberg, A., and Ryall, R. W., "Reticulo-spinal inhibition of interneurones," J. Physiol. (England) 194:225-236 (1968).

Erulkar, S. D., Sprague, J. M., Whitsel, B. L., Dogan, S., and Jannetta, P. J., "Organization of the vestibular projection to the spinal cord of the cat," J. Neurophysiol. 29:626-664 (1966).

Evarts, E. V., "Temporal patterns of discharge of pyramidal tract neurons during sleep and waking in the monkey," J. Neurophysiol. 27:152-171 (1964).

Evarts, E. V., "Relation of discharge frequency to conduction velocity in pyramidal tract neurons," J. Neurophysiol. 28:216-228 (1965).

Evarts, E. V., "Pyramidal tract activity associated with a conditioned hand movement in the monkey," J. Neurophysiol. 29:1011-1027 (1966).

Evarts, E. V., "Representation of movements and muscles by pyramidal tract neurons of the precentral motor cortex," in: Neurophysiological Basis of Normal and Abnormal Activities, New York, Hewlett, 1967, pp. 215-254.

Evarts, E. V., "Relation of pyramidal tract activity to force exerted during voluntary movement," J. Neurophysiol. 31:14-27 (1968).

Evarts, E. V., "Activity of pyramidal tract neurons during postural fixation," J. Neurophysiol. 32:375-385 (1969).

Evarts, E. V., "Activity of ventralis lateralis neurons prior to movement in the monkey," Physiologist 13:191 (1970).

Evarts, E. V., "Contrasts between activity of precentral and postcentral neurons of cerebral cortex during movement in the monkey," Brain Res. 40:25-31 (1972).

Evarts, E. V., and Thach, W. T., "Motor mechanisms of the CNS: Cerebro-cerebellar interrelations," Annu. Rev. Physiol. 31: 451-498 (1969).

Feldman, A. G., "On the functional adjustment of the nervous system during motor control or postural fixation. II. Controlled muscle parameters," Biofizika 11:498-508 (1966).

Fetz, E. E., "Pyramidal tract effects on interneurons in the cat lumbar dorsal horn," J. Neurophysiol. 31:69-80 (1968).

Fetz, E. E., and Finocchio, D. V., "Operant conditioning of isolated activity in specific muscles and precentral cells," Brain Res. 40:19-23 (1972).

Fidone, S. J., and Preston, J. B., "Patterns of motor cortex control of flexor and extensor cat fusimotor neurons," J. Neurophysiol. 32:103-115 (1969).

Foerster, O., and Altenburger, H., "Zur Physiologie und Pathophysiologie der Sehnen und Knochenphaenomene und der Dehnungsreflexe," Z. Gesamte Neurol. Psychiatr. 146:147-162 (1933).

Fox, J. L., and Kenmore, P. I., "The effect of ischemia on nerve conduction," Exp. Neurol. 17:403-419 (1967).

French, J. H., Clark, D. B., Butler, H. G., and Teasdall, R. D., "Phenylketonuria: Some observations of reflex activity," J. Pediatr. 58:17-22 (1961).

Fulton, J. F., Muscular Contraction and the Reflex Control of Movement, Baltimore, 1926.

Fulton, J. F., and Pi-Suñer, J. A., "A note concerning the probable function of various afferent end-organs in skeletal muscle," Am. J. Physiol. 83:554-562 (1928).

Gardner, E. D., and Morin, F., "Spinal pathways for projection of cutaneous and muscular afferents to the sensory and motor cortex of the monkey," Am. J. Physiol. 174:149-154 (1953).

Gasser, H. S., and Erlanger, J., Electrical Signs of Nervous Activity, Philadelphia, 1937.

Gel'fand, I. M., and Tsetlin, M. L., "On some modes of control of complex systems," Usp. Mat. Nauk 17(1):1-25 (1962).

Gel'fand, I. M., and Tsetlin, M. L., "On the mathematical modeling of mechanisms of the central nervous system," in: Models of the Structural and Functional Organization of Some Biological Systems [in Russian], Moscow, Science Publishing House, pp. 9-27, 1966.

Gel'fand, I. M., Gurfinkel', V. S., and Tsetlin, M. L., "Some considerations on the tactics of the organization of movement," Dokl. Akad. Nauk AN SSSR 139:1250-1253 (1961).

Gel'fand, I. M., Gurfinkel', V. S., and Tsetlin, M. L., "On the tactics of the control of complex systems: Physiological aspects," in: The Biological Aspects of Cybernetics [in Russian], Moscow, pp. 66-73, 1962.

Gel'fand, I. M., Gurfinkel', V. S., Kots, Ya. M., Tsetlin, M. L., and Shik, M. L., "On the synchronization of motor units and models associated with this synchronization," Biofizika 8:475-486 (1963).

Gel'fand, I. M., Gurfinkel', V. S., Tsetlin, M. L., and Shik, M. L., "Some questions on the study of movement," in: Models of the Structural and Functional Organization of Some Biological Systems [in Russian], Moscow, Science Publishing House, pp. 264-276, 1966.

Gellhorn, E., Physiological Foundations of Neurology and Psychiatry, Minneapolis, University of Minnesota Press, 1956.

Gernandt, B. E., and Gilman, S., "Generation of labyrinthine impulses, descending vestibular pathways, and modulation of vestibular activity by proprioceptive, cerebellar and reticular influences," in: Neural Mechanisms of Auditory and Vestibular Systems, Springfield, Illinois, Charles C. Thomas, 1960, pp. 324-328.

Gernandt, B. E., and Thulin, C. A., "Vestibular connections of the brain stem," Am. J. Physiol. 171:121-127 (1952).

Gernandt, B. E., and Thulin, C. A., "Vestibular mechanisms of facilitation and inhibition of cord reflexes," Am. J. Physiol. 172: 653-660 (1953).

Gernandt, B. E., and Thulin, C. A., "Reciprocal effects upon spinal motoneurons from stimulation of bulbar reticular formation," J. Neurophysiol. 18:113-129 (1955a).

Gernandt, B. E., and Thulin, C. A., "Effect of vestibular nerve section upon the spinal influence of the bulbar reticular formation," Acta Physiol. Scand. 33:120-131 (1955b).

Gernandt, B. E., Katsuki, Y., and Livingston, R. B., "Functional organization of descending vestibular influences," J. Neurophysiol. 20:453-469 (1957).

Ghez, C., and Lenzi, G. L., "Modulation of afferent transmission in the lemniscal system during voluntary movement in cat," Brain Res. 24:542 (1970).

Ghez, C., and Lenzi, G. L., "Modulation of sensory transmission in cat lemniscal system during voluntary movement," Pfluegers Arch. Gesamte Physiol. Menschen Tiere 323:273-278 (1971).

Ghez, C., and Pisa, M., "Inhibition of afferent transmission in cuneate nucleus during voluntary movement in the cat," Brain Res. 40:145-152 (1972).

Gidikov, A., The Microstructure of Voluntary Movements in Man [in Russian], Sofiya, 1970.

Gilden, L., Vaughan, H. G., and Costa, L. D., "Summated human electroencephalographic potentials associated with voluntary movements," EEG Clin. Neurophysiol. 20:433-438 (1966).

Gilson, A. S., Jr., and Mills, W. B., "Activities of single motor units in man during slight voluntary efforts," Am. J. Physiol. 133:658-669 (1941).

Gorgiladze, G. I., "The dual function of the vestibular apparatus (on the principle of the equilibrated Khegies-Bekhterev centers)," Fiziol. Zh. SSSR im. I. M. Sechenova 52:667-767 (1966).

Gottlieb, G. L., and Agarwall, G. C., "The role of the myotatic reflex in the voluntary control of movements," Brain Res. 40: 139-143 (1972).

Granit, R., Receptors and Sensory Perception: A Discussion of Aims, Means and Results of Electrophysiological Research into the Process of Reception (Silliman Memorial Lectures Series), New Haven, Connecticut, Yale University Press, 1955.

Granit, R., Basis of Motor Control, New York, Academic Press, (1970).

Granit, R., Henatsch, H.-D., and Steg, G., "Tonic and phasic ventral horn cells differentiated by post-tetanic potentiation in cat extensors," Acta Physiol. Scand. 37:114-126 (1956).

Granit, R., Pascoe, S. E., and Steg, G., "The behavior of tonic and phasic motoneurons during stimulation of recurrent collaterals," J. Physiol. (England) 138:381-400 (1957a).

Granit, R., Phillips, C. G., Skoglund, S., and Steg, G., "Differentiation of tonic from phasic alpha ventral horn cells by stretch, pinna and crossed extensor reflexes," J. Neurophysiol. 20:470-481 (1957b).

Grillner, S., and Lund, S., "The origin of a descending pathway with monosynaptic action on flexor motoneurones," Acta. Physiol. Scand. 74:274-284 (1968).

Grillner, S., Hongo, T., and Lund, S., "Interaction between the inhibitory pathways from the Deiters' nucleus and Ia afferents to flexor motoneurones," Acta Physiol. Scand. 68(277):61-79 (1966).

Grillner, S., Hongo, T., and Lund, S., "Reciprocal effects between two descending bulbospinal systems with monosynaptic connections to spinal motoneurones," Brain Res. 10:477-480 (1968).

Grillner, S., Hongo, T., and Lund, S., "Descending monosynaptic and reflex control of gamma-motoneurons," Acta Physiol. Scand. 75: 592-613 (1969).

Grillner, S., Hongo, T., and Lund, S., "The vestibulospinal tract. Effects on alpha-motoneurones in the lumbosacral spinal cord in the cat," Exp. Brain Res. 10:94-120 (1970).

Gurfinkel', V. S., and Kots, Ya. M., "Motor preadjustment in man," in: The Neural Mechanisms of Motor Activity [in Russian], Moscow, Science Publishing House, 1966, pp. 158-165.

Gurfinkel', V. S., and Pal'tsev, E. I., "Effect of the state of the segmental apparatus of the spinal cord on the performance of a simple motor reaction," Biofizika 10:855-860 (1965).

Gurfinkel', V. S., and Pal'tsev, E. I., "Reflex reaction of antagonist muscles on evocation of the tendon reflex," in: Second International Symposium on the Regulation of Movement, Sofiya, 1972, p. 18.

Gurfinkel', V. S., Ivanova, A. N., Kots, Ya. M., Pyatetskii-Shapiro, I. I., and Shik, M. L., "Quantitative characteristics of the function of motor units in the steady-state regime," Biofizika 9:636-638 (1964).

Gurfinkel', V. S., Kots, Ya. M., and Shik, M. L., Postural Regulation in Man [in Russian], Moscow, Science Publishing House, 1965.

Haase, J., and Van der Meulen, J. P., "Effects of supraspinal stimulation on Renshaw cells belonging to extensor motoneurones," J. Neurophysiol. 24:510-520 (1961).

Hagbarth, K. E., "Post-tetanic potentiation of myotatic reflexes in man," J. Neurol. Neurosurg. Psychiatry 25:1-10 (1962).

Henatsch, H.-D., and Schulte, F. J., "Reflexerregung und Eigenhemmung tonischer und phasischer Alpha-Motoneurone während chemischer Dauererregung der Muskelspindeln," Pfluegers Arch. Gesamte Physiol. Menschen Tiere 268:134-147 (1958).

Henneman, E., Somjen, G., and Carpenter, D. O., "Functional significance of cell size in spinal motoneurons," J. Neurophysiol. 28: 560-580 (1965a).

Henneman, E., Somjen, G., and Carpenter, D. O., "Excitability and inhibitability of motoneurons of different sizes," J. Neurophysiol. 28:599-620 (1965b).

Hern, J. E. C., Landgren, S., Phillips, C. G., and Porter, R., "Selective excitation of corticofugal neurones by surface-anodal stimulation of the baboon's motor cortex," J. Physiol. (England) 161:73-90 (1962).

Herrick, C. J., "Origin and evolution of the cerebellum," Arch. Neurol. Psychiatry 11:621-652 (1924).

Hess, W. R., Das Zwischenhirn: Syndrome, Lokalization, Funktionen, Basel, 1949.

Hess, W. R., The Diencephalon. Autonomic and Extrapyramidal Functions, New York, 1954.

Higgins, D. C., and Lieberman, J. S., "The muscle silent period: Variability in normal man," EEG Clin. Neurophysiol. 24:176-182 (1968).

Hines, M., "Control of movements by the cerebral cortex in primates," Biol. Rev. 18:1-31 (1943).

Hoff, A. E., Hoff, E. C., Bucy, P. C., and Pi-Suñer, J., "The production of the silent period by the synchronization of discharge of motor neurones," Am. J. Physiol. 109:123-132 (1934).

Hoff, E. C., and Hoff, A. E., "Spinal termination of the projection fibers from the motor cortex of primates," Brain 57:454-473 (1934).

Hoffmann, F. A., "Lässt sich eine posttetanische Verstärkung monosynaptischer Reflexe beim Menschen nachweisen?," Pfluegers Arch. Gesamte Physiol. Menschen Tiere 255:308-314 (1952).

Hoffmann, P., "Über die Beziehungen der Sehnenreflexe zur willkürlichen Bewegung und zum Tonus," Z. Biol. 68:351-370 (1918).

Hoffmann, P., Untersuchungen über die Eigenreflexe (Sehnenreflexe) menschlicher Muskeln, Berlin, Springer-Verlag, 1922.

Hogyes, A., "Über den Nervenmechanismus der assosierten Augenbewegungen," Monatsschr. Ohrenheilkd. 46:685-740 (1912).

Hohmann, T. C., and Goodgold, S., "A study of abnormal reflex patterns in spasticity. A new application of electrodiagnosis," Am. J. Phys. Med. 40:52-55 (1961).

Holmes, G., "The cerebellum of man," Brain 62:1-30 (1930).

Holmquist, B., and Lundberg, A., "On the organization of the supraspinal inhibitory control of interneurones of various spinal reflex arcs," Arch. Ital. Biol. 97:340-356 (1959).

Holmquist, B., and Lundberg, A., "Differential supraspinal control of synaptic actions evoked by volleys in the flexor reflex afferents in alpha-motoneurones," Acta Physiol. Scand. 54(Suppl. 186):1-51 (1961).

Homma, S., and Tateiwa, M., "Post-tetanic potentiation on spinal monosynaptic reflex in human body," J. Physiol. Soc. Jpn. 22: 1013-1020 (1960).

Hongo, T., Jankowska, E., and Lundberg, A., "Effects evoked from rubrospinal tract in cats," Experientia 21:525-526 (1965).

Hongo, T., Jankowska, E., and Lundberg, A., "Convergence of excitatory and inhibitory action on interneurones in the lumbosacral cord," Exp. Brain Res. 1:338-358 (1966).

Hongo, T., Jankowska, E., and Lundberg, A., "The rubro-spinal tract. I. Effects on alpha-motoneurones innervating hind-limb muscles in cats," Exp. Brain Res. 7:344-364 (1969a).

Hongo, T., Jankowska, E., and Lundberg, A., "The rubro-spinal tract. II. Facilitation of interneuronal transmission in reflex path to motoneurones," Exp. Brain Res. 7:365-391 (1969b).

Hongo, T., Jankowska, E., and Lundberg, A., "The rubro-spinal tract. III. Effects on primary afferent terminals," Exp. Brain Res. 15:39-53 (1972).

Hufschmidt, H.-J., "Wird die silente Periode nach direkter Muskelreizung durch die Golgi-Sehnenorgane ausgelöst?," Pfluegers Arch. Gesamte Physiol. Menschen Tiere 271:35-39 (1960).

Hufschmidt, H.-J., "The demonstration of autogenic inhibition and its significance in human voluntary movement," in: Muscular Afferents and Motor Control, Stockholm, 1966, pp. 269-274.

Hufschmidt, H.-J., and Hufschmidt, T., "Antagonist inhibition as the earliest sign of sensory-motor reaction," Nature (London) 174:607 (1954).

Hultborn, H., "Convergence on interneurons in the reciprocal Ia inhibitory pathway to motoneurones," Acta Physiol. Scand. 84, Suppl. 375 (1972).

Hultborn, H., Jankowska, E., and Lindström, S., "Inhibition in Ia inhibitory pathway by impulses in recurrent motor axon collaterals," Life Sci. 7:337-339 (1968a).

Hultborn, H., Jankowska, E., and Lindström, S., "Recurrent inhibition from motor axon collaterals in interneurones monosynaptically activated from Ia afferents," Brain Res. 9:367-369 (1968b).

Hultborn, H., Jankowska, E., and Lindström, S., "Relative contribution from different nerves to recurrent depression of Ia IPSPs in motoneurones," J. Physiol. (England) 215:637-664 (1971).

Humphrey, D. R., "Relating motor cortex spike trains to measures of motor performance," Brain Res. 40:7-18 (1972).

Humphrey, D. R., Schmidt, E. M., and Thompson, W. D., "Predicting measures of motor performance from multiple cortical spike trains," Science 170:758-762 (1970).

Hyndman, O. R., "Physiology of the spinal cord. II. The influence of chordotomy on existing motor disturbance," J. Nerv. Ment. Dis. 98:343-358 (1943).

Ioffe, M. E., "Participation of the pyramidal system in motor preadjustment in dogs," Zh. Vyssh. Nervn. Deyat. im. I. P. Pavlova 20:1292-1294 (1970).

Ioffe, M. E., "Supraspinal adjustment of the segmental apparatus prior to the performance of an instrumental movement in dogs," Zh. Vyssh. Nervn. Deyat. im. I. V. Pavlova 23:488-495 (1973).

Ioseliani, T. K., Neural Mechanisms of the Integrative Activity of the Spinal Cord, Tbilisi, "Metsniereba," 1970.

Jackson, J. H. "On the anatomical and physiological localization of movements in the brain," Lancet 1:162 (1873). Reprinted in Selected writings (ed. Taylor, J.) 1:68 (1931).

Jang, R., and Hassler, R., "The extrapyramidal motor system," in: Handbook of Physiology, Sec. 2, Pt. 2, Washington, American Physiology Society, 1960, pp. 863-927.

Jansen, J. K. S., and Matthews, P. B. C., "The effects of fusimotor activity on the static responsiveness of primary and secondary endings of muscle spindles in the decerebrate cat," Acta Physiol. Scand. 55:376-386 (1962a).

Jansen, J. K. S., and Matthews, P. B. C., "The central control of the dynamic response of muscle spindle receptors," J. Physiol. (England) 161:357-378 (1962b).

Jansen, J. K. S., and Rudjord, T., "On the silent period and Golgi tendon organs of the soleus muscle of the cat," Acta Physiol. Scand. 62:364-379 (1964).

Jasper, H. H., and Bertrand, G., "Thalamic units involved in somatic sensation and voluntary and involuntary movements in man," in: The Thalamus, New York, Columbia University Press, 1966, pp. 365-390.

Jasper, H., Richi, G., and Down, B., "Microelectrode analysis of cortical cell discharge during the formation of conditioned defense reflexes in monkeys," in: Electroencephalographic Investigation of Higher Nervous Activity [in Russian], Moscow, Publishing House of the Academy of Sciences of the USSR, 1962, pp. 129-146.

Job, C., "Über autogene Inhibition und Reflexumkehr bei spinalisierten und decerebrierten Katzen," Pfluegers Arch. Gesamte Physiol. Menschen Tiere 256:406-418 (1953).

Jones, S. G., and Powell, T. P. S., "Connexions of the somatic sensory cortex of the Rhesus monkey. I. Ipsilateral connexions," Brain 92:477-502 (1969).

Kalinovskaya, I. Ya., and Yusevich, Yu. S., "On the question of oculographic and electromyographic investigation of changes in muscle tonus on vestibular stimulation," Zh. Nevropatol. Psikhiatr. im. S. S. Korsakova 63:668-672 (1963).

Kas'yanov, V. M., "The characteristics of positional excitation in human locomotor reactions," Byull. Eksp. Biol. Med. 30(7): 16-20 (1950).

Kato, M., Takamura, H., and Fujimori, B., "Studies on effects of pyramid stimulation upon flexor and extensor motoneurones and gamma-motoneurones," Jpn. J. Physiol. 14:34-44 (1964).

Kemp, J. M., and Powell, T. P. S., "The cortico-striate projection in the monkey," Brain 93:525-546 (1970).

Kernell, D., "The adaptation and the relation between discharge frequency and current strength of cat lumbosacral motoneurones stimulated by long-lasting injected currents," Acta Physiol. Scand. 65:65-73 (1965a).

Kernell, D., "The limits of firing frequency in cat lumbosacral motoneurones posessing different time course of after-hyper-polarization," Acta Physiol. Scand. 65:87-100 (1965b).

Khazovskaya, E. G., "Dynamics of the monosynaptic reflex (H-reflex) during changes in afferentation from the posterior tibial muscle group in man," in: The Structure and Function of the Nervous System [in Russian], Moscow, Medicine Publishing House, 1965, pp. 123-130.

Knapp, H. D., Taub, E., and Berman, A. J., "Movements in monkeys with deafferented forelimbs," Exp. Neurol. 7:305-315 (1963).

Koeze, T. H., "The independence of corticomotoneuronal and fusimotor pathways in the production of muscle contraction by motor cortex stimulation," J. Physiol. (England) 197:87-105 (1968).

Koeze, T. H., Phillips, C. G., and Sheridan, J. D., "Thresholds of cortical activation of muscle spindles and alpha-motoneurons of the baboon's hand," J. Physiol. (England) 195:419-449 (1968).

Koizumi, K., Ushiyama, J., and Brooks, C. M., "A study of reticular formation action of spinal interneurons and motoneurons," Jpn. J. Physiol. 9:282-303 (1959).

Kolodnaya, A. Ya., "Changes in the electrical activity of muscles during development of differentiated motor reactions in man," in: Questions of the Study of Higher Neurodynamics As Related to the Problems of Psychology [in Russian], Moscow, 1957, pp. 168-185.

Kornhüber, H. H., and Deecke, L., "Hirnpotentialänderungen beim Menschen vor und nach Willkürbewegungen, dargestellt mit

Magnetbandspeicherung und Rückwärtsanalyse," Pfluegers Arch. Gesamte Physiol. Menschen Tiere 281:52-64 (1964).

Kornhüber, H. H., and Deecke, L., "Hirnpotentialänderungen bei Willkürbewegungen und passiven Bewegungen des Menschen: Bereitschaftspotential und reafferente Potentiale," Pfluegers Arch. Gesamte Physiol. Menschen Tiere 284:1-17 (1965).

Korobkov, A. V., Ovsyannikov, A. V., and Khomyakova, G. D., "Supraspinal inhibition of intersegmental connections during the formation of an acyclic movement," in: The Processing of Visual Information and the Regulation of Motor Activity [in Russian], Sofiya, 1969, p. 34.

Kosilov, S. A., and Farfel', V. S., "Effect of working rhythm on the development of a stroke during hammering movements," Uch. Zap. Leningr. Gos. Univ. Ser. Biol. Nauk, No. 23, 247-256 (1938).

Kostyuk, P. G., "Post-tetanic changes in the reflex reactions of cord motor cells," in: Questions of Physiology [in Russian], Vol. 10, Kiev, pp. 58-71 (1954).

Kostyuk, P. G., The Two-Neuron Reflex Arc [in Russian], Moscow, State Medical Publishing House, 1959.

Kostyuk, P. G., "Neuronal mechanisms of corticospinal motor systems," Fiziol. Zh. SSSR im. I. M. Sechenova 53:1311-1321 (1967).

Kostyuk, P. G., "Investigation of the spinal neuronal mechanisms of motor control systems," Neirofiziologiya 2:189-202 (1970).

Kostyuk, P. G., Structure and Function of Descending Cord Systems [in Russian], Leningrad, Science Publishing House, 1973.

Kostyuk, P. G., and Pilyavskii, A. I., "Post-synaptic potentials of spinal motoneurons in the presence of rubrospinal actions," Zh. Vyssh. Nervn. Deyat. im. I. P. Pavlova 17:497-504 (1967).

Kostyuk, P. G., and Pilyavskii, A. I., "Synaptic processes in intermediate cord neurons in the presence of rubrospinal actions," Neirofiziologiya 1:158-166 (1969).

Kostyuk, P. G., and Preobrazhenskii, N. N., "Characteristics of the inhibition of polysynaptic reflex reactions during stimulation of medullar formation," Fiziol. Zh. SSSR im. I. M. Sechenova 53:1048-1055 (1967).

Kostyuk, P. G., and Vasilenko, D. A., "The processing of a cortical motor signal in the spinal cord," in: Investigations of Neural Elements and Systems [in Russian], Moscow, World Publishing House, 1968, pp. 172-181.

Kostyuk, P. G., Vasilenko, D. A., and Zadorozhnyi, A. G., "Lumbar motoneuron reactions evoked by the activity of propriospinal pathways," Neirofiziologiya 1:5-14 (1969).

Kots, Ya. M., "Analysis of the change in the state of the segmental cord apparatus in man during voluntary movements," in: Xth Congress of the I. P. Pavlov All-Union Physiology Society, Vol. 2, Issue 1, Erevan, 1964, p. 422.

Kots, Ya. M., "Afferentation of the 'command,'" in: Materials of the IXth All-Union Scientific Conference on Physiology, Morphology, Biochemistry and Biomechanics of Muscular Activity, Vol. 2, pp. 47-48, 1966.

Kots, Ya. M., "The reflex excitability of human special motoneurons under conditions of transient ischemic deafferentation," Byull. Eksp. Biol. Med. 65(9):32-37 (1968).

Kots, Ya. M., "On the supraspinal control of the segmental centers of antagonist muscles in man. I. Reflex excitability of the motoneurons of antagonist muscles during the organization of voluntary movement," Biofizika 14:167-172 (1969a).

Kots, Ya. M., "On the supraspinal control of the segmental centers of antagonist muscles in man. II. The reflex excitability of the motoneurons of antagonist muscles during the organization of their sequential activity," Biofizika 14:1087-1094 (1969b).

Kots, Ya. M., "Spinal 'adjustment' prior to voluntary phasic movement performed after static effort," Zh. Vyssh. Nervn. Deyat. im. I. P. Pavlova 19:363-365 (1969c).

Kots, Ya. M., "Investigation of the participation of the gamma loop in the organization of voluntary movement," Zh. Vyssh. Nervn. Deyat. im. I. P. Pavlova 19:862-869 (1969d).

Kots, Ya. M., "The organization of voluntary movement under conditions of temporary ischemic deafferentation," in: Problems of the Physiology of the Motor Apparatus [in Russian], Moscow, 1970, pp. 69-89.

Kots, Ya. M., "The interaction of cortical and vestibular motor centers during the organization of voluntary movement," in: The Physiology and Pathophysiology of the Limbicoreticular System [in Russian], Moscow, Science Publishing House, 1971, pp. 119-124.

Kots, Ya. M., "Spinal mechanisms of the organization of voluntary movement" (Doctoral dissertation), Moscow, 1972.

Kots, Ya. M., and Krinskii, V. I., "The monosynaptic H-reflex in man recorded in the soleus and medial gastrocnemius muscles during rest," Fiziol. Zh. SSSR im. I. M. Sechenova 53:784-790 (1967).

Kots, Ya. M., and Krinskii, V. I., "The monosynaptic H-reflex in man recorded in the soleus and medial gastrocnemius muscles under conditions of facilitation," Fiziol. Zh. SSSR im. I. M. Sechenova 54:17-23 (1968).

Kots, Ya. M., and Mart'yanov, V. A., "The application of the monosynaptic H-reflex method in recording the effects of electrical stimulation of the human vestibular apparatus," Kosm. Biol. Med., No. 3, 89-93 (1967).

Kots, Ya. M., and Mart'yanov, V. A., "The blockade of vestibulospinal influences during the organization of voluntary movement," Biofizika 13:834-842 (1968).

Kots, Ya. M., and Naidin, V. L., "On the role of kinesthetic sensation in the control of voluntary movement," Vopr. Psikhol., No. 5, 114-122 (1966).

Kots, Ya. M., and Zaitsev, A. A., "The differentiation of 'slow' and 'fast' motoneurons in man by the effects of posttetanic potentiation and ischemic deafferentation," Fiziol. Zh. SSSR im. I. M. Sechenova 56:55-63 (1970).

Kots, Ya. M., and Zhukov, V. I., "On the supraspinal control of the segmental centers of antagonist muscles in man. 3. The 'adjustment' of the spinal reciprocal inhibitory apparatus during the organization of voluntary movement," Biofizika 16:1093-1099 (1971).

Kots, Ya. M., Krinskii, V. I., Naidin, V. L., and Shik, M. L., "The control of joint movements and kinesthetic afferentation," in: Models of the Structural and Functional Organization of Some Biological Systems [in Russian], Moscow, Science Publishing House, 1966, pp. 302-309.

Koz'myan, E. I., "The relationship of excitation and inhibition times of antagonist muscles," Zh. Vyssh. Nervn. Deyat. im. I. P. Pavlova 17:125-133 (1967).

Koz'myan, E. I., "On the relationship of the activation times of antagonist muscles of man on the disruption of an experimentally developed stereotype," Zh. Vyssh. Nervn. Deyat. im. I. P. Pavlova 18:514-516 (1968).

Kryzhanovskii, G. N., "Some characteristics of the integrative activity of the spinal reflex apparatus under conditions of the disruption of inhibitory mechanisms," in: The Integrative Activity of the Nervous System in Normal Subjects and in the Presence of Pathology [in Russian], Moscow, Medicine Publishing House, 1968, pp. 21-35.

Kubota, K., Iwamura, Y., and Niimi, Y., "Monosynaptic reflex and natural sleep in the cat," J. Neurophysiol. 28:125-134 (1965).

Kugelberg, E., and Hagbarth, K., "Spinal mechanism of the abdominal and erector spinal skin reflexes," Brain 18:290-318 (1958).

Kukinova, L. P., and Ivanova, M. P., "The slow negative wave in the human EEG and reaction time," Fiziol. Zh. SSSR im. I. M. Sechenova 60:981-985 (1974).

Kuno, M., "Excitability following antidromic activation in spinal motoneurones supplying red muscles," J. Physiol. (England) 149: 374-393 (1959).

Kuypers, H. G. J. M., "Some projections from the pericentral cortex to the pons and lower brain stem in monkey and chimpanzee," J. Comp. Neurol. 110:221-225 (1958a).

Kuypers, H. G. J. M., "Cortico-bulbar connexions to the pons and lower brain stem in man," Brain 81:364-388 (1958b).

Kuypers, H. G. J. M., "Central cortical projections to motor and somatosensory cell groups. An experimental study in the Rhesus monkey," Brain 83:161-184 (1960).

Kuypers, H. G. J. M., "The descending pathways to the spinal cord, their anatomy and function," Prog. Brain Res. 11:178-200 (1964).

Ladpli, R., and Brodal, A., "Experimental studies of commissural and reticular formation projections from the vestibular nuclei in the cat," Brain Res. 8:65-96 (1968).

Lance, J. W., De Gail, P., and Neilson, P. D., "Tonic and phasic spinal cord mechanisms in man," J. Neurol. Neurosurg. Psychiatry 29:535-544 (1966).

Landgren, S., Phillips, C. G., and Porter, R., "Minimal synaptic actions of pyramidal impulses on some alpha-motoneurones of the baboon's hand and forearm," J. Physiol. (England) 161:91-111 (1962a).

Landgren, S., Phillips, C., and Porter, R., "Cortical fields of origin of the monosynaptic pyramidal pathways to some alpha-motoneurones of the baboon's hand and forearm," J. Physiol. (England) 161:112-125 (1962b).

Laporte, Y., and Bessou, P., "Modifications d'excitabilité de motoneurones homonymes provoquées par l'activation physiologique de fibres afférentes d'origine musculaire du groupe II," J. Physiol. (France) 51:897-908 (1959).

Lassek, A. M., The Pyramidal Tract. Its Status in Medicine, Springfield, Illinois, Charles C. Thomas, 1954.

Laursen, A. M., and Wiesendanger, M., "Pyramidal effect on alpha and gamma motoneurons," Acta Physiol. Scand. 67:165-172 (1966).

Lavy, A. H., and Valbo, A. B., "Motoneuron activation by low-intensity tetanic stimulation of muscle afferents in man," Exp. Neurol. 18:383-391 (1967).

Lawrence, D. G., and Kuypers, G. J. M., "The functional organization of the motor system in the monkey. I. The effect of bilateral pyramidal lesions," Brain 91:1-14 (1968a).

Lawrence, D. G., and Kuypers, G. J. M., "The functional organization of the motor system in the monkey. II. The effects of lesions of the descending brain-stem pathways," Brain 91:15-36 (1968b).

Leshchinyuk, I. I., "Investigation of the functional organization of dorsal horn neurons," Zh. Vyssh. Nervn. Deyat. im. I. P. Pavlova 18:126-132 (1968).

Levy, R., "The relative importance of the gastrocnemius and soleus muscles in the ankle jerk of man," J. Neurol. Neurosurg. Psychiatry 26:148-150 (1963).

Lewy, H., Die Lehre vom Tonus und Bewegung, Berlin, 1923.

Limanskii, Yu. P., and Preobrazhenskii, N. N., "The medulla oblongata, pons, and midbrain," in: The General and Specific Physiology of the Nervous System [in Russian], Leningrad, Science Publishing House, 1969, pp. 255-287.

Liu, C. N., and Chambers, W. W., "An experimental study of the corticospinal system in the monkey (<u>Macaca mulatta</u>)," J. Comp. Neurol. 123:257-284 (1964).

Livanov, M. N., and Raeva, S. N., "The microelectrode investigation of the human brain," Dokl. Akad. Nauk SSSR 204:507-509 (1972).

Lloyd, D. P. C., "A direct central inhibitory action of dromically conducted impulses," J. Neurophysiol. 4:184-190 (1941a).

Lloyd, D. P. C., "The spinal mechanism of the pyramidal system in cats," J. Neurophysiol. 4:525-546 (1941b).

Lloyd, D. P. C., "Mediation of descending long spinal reflex activity," J. Neurophysiol. 5:435-458 (1942).

Lloyd, D. P. C., "Reflex action in relation to pattern and peripheral source of afferent stimulation," J. Neurophysiol. 6:112-119 (1943a).

Lloyd, D. P. C., "The interaction of antidromic and orthodromic volleys in a segmental spinal motor nucleus," J. Neurophysiol. 6:143-152 (1943b).

Lloyd, D. P. C., "Facilitation and inhibition of spinal motoneurons," J. Neurophysiol. 9:421-438 (1946).

Lloyd, D. P. C., "Post-tetanic potentiation of response in monosynaptic reflex pathways of the spinal cord," J. Gen. Physiol. 33:147-170 (1949).

Lloyd, D., and McIntyre, A., "Analysis of forelimb-hindlimb reflex activity in acutely decapitated cats," J. Neurophysiol. 11:455-470 (1948).

Lorento-de-No, R., "Ausgewählte Kapitel aus der verleichenden Physiologie des Labyrinthes," Ergeb. Physiol. 32:73-242 (1931).

Lund, S., and Pompeiano, O., "Monosynaptic excitation of alphamotoneurons from supraspinal structures in the cat," Acta Physiol. Scand. 73:1-21 (1968).

Lundberg, A., "Supraspinal control of transmission in reflex paths to motoneurones and primary afferents," Prog. Brain Res. 12:197-219 (1964).

Lundberg, A., "Integration in the reflex pathway," in: Muscular Afferents and Motor Control, Stockholm, 1966, pp. 275-305.

Lundberg, A., "The excitatory control of the I-a inhibitory pathway," in: Excitatory Synaptic Mechanisms, Oslo, University Press, 1970, pp. 333-340.

Lundberg, A., and Voorhoeve, P., "Effects from the pyramidal tract on spinal reflex arcs," Acta Physiol. Scand. 56:201-219 (1962).

Lundberg, A., and Vyklicky, L., "Brain stem control of reflex paths to primary afferents," Acta Physiol. Scand. 59(Suppl. 213): 91 (1963).

Lundberg, A., and Winsbury, G., "Selective adequate activation of large afferents from muscle spindles and Golgi tendon organs," Acta Physiol. Scand. 49:155-164 (1960).

Lundberg, A., Norrsell, U., and Voorhoeve, P., "Pyramidal effects on lumbosacral interneurones activated by somatic afferents," Acta Physiol. Scand. 56:220-229 (1962).

Luriya, A. R., The Higher Cortical Functions in Man and Their Distruption by Localized Lesions of the Brain [in Russian], Moscow State University Press, 1962.

Luriya, A. R., The Human Brain and Psychic Processes [in Russian], Moscow, Pedagogy Publishing House, 1970.

Luschei, E., Saslow, C., and Glickstein, M., "Muscle potentials in reaction time," Exp. Neurol. 18:429-442 (1967).

Luschei, E. S., Johnson, R. A., and Glickstein, M., "Response of neurons in the motor cortex during performance of a simple repetitive arm movement," Nature (London) 217:190-191 (1968).

Luschei, E. S., Garthwaite, C. R., and Armstrong, M. E., "Relationship of firing patterns of units in face area of monkey precentral cortex to conditioned jaw movements," J. Neurophysiol. 34:552-561 (1971).

MacLean, K. B., and Leffman, H., "Supraspinal control of Renshaw cells," Exp. Neurol. 18:94-104 (1967).

Magladery, J. W., "Some observations on spinal reflexes in man," Pfluegers Arch. Gesamte Physiol. Menschen Tiere 261:302-321 (1955).

Magladery, J. W., and McDougal, D. B., "Electrophysiological studies of nerve and reflex activity in normal man. I. Identification of certain reflexes in the EMG and the conduction velocity of peripheral nerve fibers," Bull. Johns Hopkins Hosp. 86:265-290 (1950).

Magladery, J. W., McDougal, D. B., and Stoll, J., "Electrophysiological studies of nerve and reflex activity in normal man. II. The effect of peripheral ischemia," Bull. Johns Hopkins Hosp. 86:291-312 (1950).

Magladery, J. W., Porter, W. E., Park, A. M., and Teasdall, R. D., "Electrophysiological studies of nerve and reflex activity in normal man. IV. The two-neurone reflex and identification of certain action potentials from spinal roots and cord," Bull. Johns Hopkins Hosp. 88:499-519 (1951a).

Magladery, J. W., Teasdall, R. D., Park, A. M., and Porter, W. E., "Electrophysiological studies of nerve and reflex activity in normal man. V. Excitation and inhibition of two-neurone reflexes by afferent impulses in the same nerve trunk," Bull. Johns Hopkins Hosp. 88:520-537 (1951b).

Magladery, J. W., Teasdall, R. D., Park, A. M., and Languth, H. W., "Electrophysiological studies of nerve and reflex activity in patients with lesions of the nervous system. I. A comparison of motoneurone excitability following afferent nerve volleys in normal persons and patients with upper motor neurone lesions," Bull. Johns Hopkins Hosp. 91:219-244 (1952).

Magnus, R. (1924), Regulation of Body Processes [in Russian], Moscow, Leningrad, Publishing House of the Academy of Science of the USSR, 1962.

Magoun, H. W., The Waking Brain, Springfield, Illinois, Charles C. Thomas, 1963.

Makarova, L. A., "The reflex excitability of functionally diverse spinal motoneurons in the presence of spastic hemiparesis" (Master's thesis), Leningrad, 1973.

Maksimova, E. V., "The activation of gamma-motoneurons by pyramidal impulsation," in: Mechanisms of the Descending Control of Spinal Cord Activity [in Russian], Leningrad, Science Publishing House, 1971, pp. 87-90.

Maksimova, E. V., and Sverdlov, S. M., "On the influence of pyramidal impulses on the motor nuclei of the spinal cord," in: Neural Mechanisms of Motor Activity [in Russian], Moscow, Science Publishing House, 1966, pp. 115-129.

Malis, L. I., Pribram, K. H., and Kruger, L., "Action potentials in 'motor' cortex evoked by peripheral nerve stimulation," J. Neurophysiol. 16:161-167 (1953).

Mart'yanov, V. A., "The time of the spinal component of a motor reaction in man" (Master's thesis), Moscow, 1968.

Mart'yanov, V. A., and Kopylov, Yu. A., "The nature of the reflex response of the motoneurons of human hand muscles," Zh. Vyssh. Nervn. Deyat. im. I. P. Pavlova 21:1092-1095 (1971).

Maruhashi, J., and Wright, E. B., "Effect of oxygen lack on the single isolated mammalian (rat) nerve fiber," J. Neurophysiol. 30:434-453 (1967).

Matthews, B. H. C., "Nerve endings in mammalian muscle," J. Physiol. (England) 78:1-53 (1933).

Matthews, P. B. C., "Muscle spindles and their motor control," Physiol. Rev. 44:219-288 (1964).

Mayer, R. F., and Mawdsley, C., "Studies in man and cat of the significance of the H-wave," J. Neurol. Neurosurg. Psychiatry 28:201-211 (1965).

McDonald, W. I., "The effects of experimental demyelination on conduction in peripheral nerve: A histological and electrophysiological study. II. Electrophysiological observations," Brain 86:501-524 (1963).

McLeod, J. G., "H-reflex studies in patients with cerebellar disorders," J. Neurol. Neurosurg. Psychiatry 32:21-27 (1969).

Megirian, D., and Troth, R., "Vestibular and muscle nerve connections of pyramidal tract neurons of cat," J. Neurophysiol. 27: 481-492 (1964).

Merton, P. A., "Significance of the 'silent period' of muscles," Nature (London) 166:733-734 (1950).

Merton, P. A., "The silent period in a muscle of the human hand," J. Physiol. (England) 114:183-198 (1951).

Merton, P. A., "Speculations on servocontrol of movement," in: The Spinal Cord, Ciba Symposium, London, Churchill, 1953, pp. 247-260.

Mizuno, Y., Tanaka, R., and Yanagisawa, N., "Reciprocal group I inhibition on triceps surae motoneurons in man," J. Neurophysiol. 34:1010-1017 (1971).

Mortimer, E. M., and Akert, K., "Cortical control and representation of fusimotor neurons," Am. J. Phys. Med. 40:228-248 (1961).

Munk, H., Über die Funktionen von Hirn und Rückenmark. Gesammelte Mitteilungen, Neue Folge, Berlin, 1909.

Naidel', A. V., and Pal'tsev, E. I., "On the 'adjustment' of the segmental apparatus of the spinal cord in man during a conditioned reaction," Zh. Vyssh. Nervn. Deyat. im. I. P. Pavlova 15:940-942 (1965).

Nakagoshi, J., "Certain influences of the cerebellum on impulses to and from the somato-sensory cortex in the cat," J. Kyoto Prefect. Med. Univ. 75:1-14 (1966).

Nakayama, T., and Hori, T., "Tonic and kinetic components of the evoked electromyogram," Jpn. J. Physiol. 17:415-428 (1967).

Nauta, W. J. H., and Mehler, W. R., "Projections of the lentiform nucleus in the monkey," Brain Res. 1:3-42 (1966).

Nebylitsyn, V. D., and Bazilevich, T. F., "Evoked motor cortex potentials in man," Fiziol. Zh. SSSR im. I. M. Sechenova 56: 1682-1691 (1970).

Nesmeyanova, T. N., The Stimulation of Restorative Processes in the Presence of Cord Trauma [in Russian], Moscow, Science Publishing House, 1971.

Norton, A. C., "The dorsal column system of the spinal cord. Its anatomy, physiology, phylogeny and sensory function," in: U.C.L.A. Brain Information Service, Los Angeles, 1969, pp. 62-80.

Nyberg-Hansen, R., "Origin and termination of fibers from the vestibular nuclei descending in the medial longitudinal fasciculus," J. Comp. Neurol. 122:355-376 (1964).

Nyberg-Hansen, R., "Functional organization of descending supraspinal fiber systems to the spinal cord. Anatomical observations and physiological correlations," Ergeb. Anat. Entwicklungsgeschich. 39:3-48 (1966).

O'Keefe, J., and Gaffan, D., "Response properties of units in the dorsal column nuclei of the freely moving rat. Changes as a function of behavior," Brain Res. 31:374-375 (1971).

Orbeli, L. A., "Lectures on the physiology of the nervous system," Leningrad-Moscow, OGIZ, 1934.

Orlov, I. V., "Materials on the electrophysiological characteristics of the vestibular apparatus of birds," Fiziol. Zh. SSSR im. I. M. Sechenova 48:24-30 (1962).

Ovsyannikov, A. V., "The effect of dorsal root deafferentation on the execution of precise instrumental motor reflexes," in: The Organization of Interneuronal Connections [in Russian], Moscow, 1967, pp. 77-86.

Ovsyannikov, A. V., and Khomyakova, G. D., "Motor preadjustment in man under conditions of choice," Zh. Vyssh. Nervn. Deyat. im. I. P. Pavlova 19:525-527 (1969).

Paillard, J., "Functional organization of afferent innervation of muscle studies in man by monosynaptic testing," Am. J. Phys. Med. 38:239-247 (1959).

Paillard, J., "The patterning of skilled movements," in: Handbook of Physiology, Sect. 1, Pt. 3, Washington, American Physiology Society, 1960, pp. 1679-1708.

Paintal, A. S., "Facilitation and depression of muscle stretch receptors by repetitive antidromic stimulation, adrenaline and asphyxia," J. Physiol. (England) 148:252-266 (1959).

Patton, H. D., and Amassian, V. E., "The pyramidal tract: Its activation and functions," in: Handbook of Physiology, Sect. 1, Pt. 2 (J. Field et al., eds.) Washington, American Physiology Society, 1960, pp. 837-861.

Peimer, N. A., and Perli, P. D., "On the analysis of normal and pathological variants of the patellar reflex," in: Problems in Electrophysiology [in Russian], Leningrad, 1950, pp. 137-142.

Person, R. S., Antagonist Muscles in Human Movements [in Russian], Moscow, Science Publishing House, 1965.

Phillips, C. G., "Changing concepts of the precentral motor area," in: Brain and Conscious Experience (J. C. Eccles, ed.), New York, Springer-Verlag, 1966, pp. 389-421.

Phillips, C. G., and Porter, R., "The pyramidal projection to motoneurones of some muscle groups of the baboon's forelimb," Prog. Brain Res. 12:222-242 (1964).

Pierrot-Deseilligny, E., Lacert, P., and Cathala, H. P., "Amplitude et variabilité des reflexes monosynaptiques avant un mouvement volontaire," Physiol. Behav. 7:495-508 (1971).

Pilyavskii, A. I., and Skibo, G. G., "Connection of the rubrospinal tract with different groups of lumbar cord neurons," Byull. Eksp. Biol. Med. 68(9):3-7 (1969).

Pompeiano, O., "Analisi degli effetti della stimolazione electrica del nucleo rosso nel gatto decerebrato," Rend. Accad. Naz. Lincei Cl. Sci. Fis. Mat. Natur. 22:100-103 (1957).

Pompeiano, O., "Organizzazione somatotopica delle risposte posturali alla stimulazione elettrica del nucleo di Deiters nel gatto decerebrato," Arch Sci. Biol. 44:497-511 (1960).

Pompeiano, O., Diete-Spiff, K., and Carli, G., "Two pathways transmitting vestibulo-spinal influences from the lateral vestibular nucleus of Deiters of extensor fusimotor neurones," Pfluegers Arch Gesamte Physiol. Menschen Tiere 293:272-275 (1967).

Popov, S. V., "Effect of suprasegmental disorders on voluntary and reflex motor acts in the child" (Doctoral dissertation), Leningrad, 1972.

Poppele, R. E., "Response of gamma and alpha motor systems to phasic and tonic vestibular inputs," Brain Res. 6:535-547 (1967).

Porter, R., "Relationship of the discharges of cortical neurones to movement in free-to-move monkeys," Brain Res. 40:39-43 (1972).

Porter, R., and Muir, R. B., "The meaning for motoneurones of the temporal pattern of natural activity in pyramidal tract neurones of conscious monkeys," Brain Res. 34:127-142 (1971).

Porter, R., Lewis, M., and Horne, M., "Analysis of patterns of natural activity of neurones in the precentral gyrus of conscious monkeys," Brain Res. 34:99-113 (1971).

Predtechenskaya, K. S., "On the posttetanic change in the reflex responses of the ventral root during rhythmic stimulation of the afferent nerve," Fiziol. Zh. SSSR im. I. M. Sechenova 51: 749-754 (1965).

Preston, J. B., and Whitlock, D. G., "Precentral facilitation and inhibition of spinal motoneurones," J. Neurophysiol. 23:154-170 (1960).

Preston, J. B., and Whitlock, D. G., "Intracellular potentials recorded from motoneurons following precentral gyrus stimulation in the primate," J. Neurophysiol. 24:91-100 (1961).

Preston, J. B., and Whitlock, D. G., "A comparison of motor cortex effects on slow and fast muscle innervations in the monkey," Exp. Neurol. 7:327-341 (1963).

Preston, J. B., Shende, M. C., and Uemura, K., "The motor cortex-pyramidal system: Patterns of facilitation and inhibition on motoneurons innervating limb musculature of cat and baboon and their possible adaptive significance," in: Neurophysiological Basis of Normal and Abnormal Motor Activities (M. D. Yahr and D. P. Purpura, eds.), New York, Raven Press, 1967, pp. 61-74.

Pyatetskii-Shapiro, I. I., and Shik, M. L., "On the question of the spinal regulation of movements," Biofizika 9:488-492 (1964).

Renshaw, B., "Reflex discharges in branches of the crural nerve," J. Neurophysiol. 5:487-498 (1942).

Rexed, B., "Some aspects of the cytoarchitectonics and synaptology of the spinal cord," Prog. Brain Res. 11:52-92 (1964).

Roitbak, A. I., and Dedabrishvili, Ts. I., "EEG changes during muscular activity," in: Sports Medicine Proceedings of the XIIth Anniversary of the International Congress of Sports Medicine [in Russian], Moscow, 1958, p. 607.

Rokotova, N. A., Bogina, I. D., Gorbunova, I. M., Pavlov, V. N., and Rogovenko, E. S., "The temporal and spatial organization of human movement," in: The Perception of Time and Space [in Russian], Leningrad, Science Publishing House, 1969, pp. 114-116.

Rokotova, N. A., Berezhnaya, E. K., Bogina, I. D., Gorbunova, I. M., and Rogovenko, E. S., Motor Tasks and Executive Activity. An Investigation of Coordinated Hand Movements [in Russian], Leningrad, Science Publishing House, 1971.

Romanes, G. J., "The motor pools of the spinal cord," Prog. Brain Res. 11:93-116 (1964).

Rossi, J. F., and Zanchetti, A., The Reticular Formation of the Brain Stem [in Russian], Moscow, Foreign Literature Publishing House, 1960.

Rusinov, V. S., Dominance: An Electrophysiological Investigation [in Russian], Moscow, Medicine Publishing House, 1969.

Rusinov, V. S., and Chugunov, S. A., "The patellar reflex and disruptive inhibition," Byull. Eksp. Biol. Med. 20(6):33-36 (1945).

Russell, J. R., and Myer, W. de, "The quantitative origin of pyramidal axons of _Macaca rhesus_, with some remarks on the slow rate of axolysis," Neurology 11:96-108 (1961).

Sadjadpour, K., and Brodal, A., "The vestibular nuclei in man. A morphological study in the light of experimental findings in the cat," J. Hirnforsch. 10:299-323 (1968).

Sasaki, K., and Tanaka, T., "Phasic and tonic innervation of spinal alpha motoneurons from upper brain centers," Jpn. J. Physiol. 14:56-66 (1964).

Sasaki, K., Nawikawa, A., and Hashiramoto, S., "Effects of stimulation of the pyramidal tract and striate body upon spinal motoneurons," Jpn. J. Physiol. 10:403-413 (1960).

Scheibel, M. E. and Scheibel, A. B., "Terminal patterns in cat spinal cord. III. Primary afferent collaterals," Brain Res. 13: 417-443 (1969).

Schlegel, H.-J., and Sontag, K.-H., "Reflektorische Aktivierung prätibialer Fusimotoneurone der Katze durch Reizung niedrigschwelliger antagonistischer Muskelafferenzen," Pfluegers Arch. Gesamte Physiol. Menschen Tiere 319:200-204 (1970).

Schmidt, R. F., and Willis, W. D., "Depolarization of central terminals of afferent fibers in the cervical spinal cord of the cat," J. Neurophysiol. 26:44-60 (1963).

Schoen, J. H. R., "Untersuchungen über die Dauer der Innervationstille (silent period) nach monosynaptischen Reflexen (Eigen-Reflexen) und nach antidromer Reizung," Pfluegers Arch. Gesamte Physiol. Menschen Tiere 254:205-213 (1951).

Schoen, J. H. R., "Comparative aspects of the descending fibre systems in the spinal cord," Prog. Brain Res. 11:203-222 (1964).

Severin, F. V., "The role of the gamma motor system in the performance of movements elicited by natural stimulation of supraspinal centers," Fiziol. Zh. SSSR im. I. M. Sechenova 51:453-457 (1966).

Severin, F. V., Shik, M. L., and Orlovskii, G. N., "The work of muscles and single motoneurons during control locomotion," Biofizika 12:660-668 (1967).

Severin, F. V., Orlovskii, G. N., and Shik, M. L., "Recurrent influences on the work of single motoneurons during controlled locomotion," Byull. Eksp. Biol. Med. 66(7):5-9 (1968).

Shapovalov, A. I., The Cellular Mechanisms of Synaptic Transmission [in Russian], Moscow, Medicine Publishing House, 1966.

Shapovalov, A. I., "Extrapyramidal synaptic influences on spinal neurons," in: Synaptic Processes [in Russian], Kiev, Naukova Dumka," 1968, pp. 215-227.

Shapovalov, A. I., "The monosynaptic control of spinal motoneurons by various levels of the brain," Neirofiziologiya 2:203-215 (1970).

Shapovalov, A. I., "The evolution of neuronal systems of suprasegmental motor control (a survey)," Neirofiziologiya 4:453-470 (1972).

Shapovalov, A. I., and Karamyan, O. A., "Short-latency interstitial and rubrospinal synaptic influences on alpha-motoneurons," Byull. Eksp. Biol. Med. 66(12):3-7 (1968).

Shapovalov, A. I., and Saf'yants, V. I., "The potentiation of suprasegmental synaptic influences on alpha-motoneurons," Fiziol. Zh. SSSR im. I. M. Sechenova 54:1262-1269 (1968).

Shapovalov, A. I., and Shapovalova, K. B., "Alpha-motoneuron activity on rhythmic stimulation of red nucleus and the action of strychnine on rubrospinal effects," Dokl. Akad. Nauk SSSR 168:1430-1433 (1966).

Shapovalov, A. I., Kurchavyi, G. G., and Stroganova, M. L., "Synaptic mechanisms of vestibulospinal influences on alpha-motoneurons," Fiziol. Zh. SSSR im. I. M. Sechenova 52:1401-1409 (1966).

Shapovalov, A. I., Grantyn', A. A., and Kurchavyi, G. G., "Short-latency reticulospinal projections to alpha-motoneurons," Byull. Eksp. Biol. Med. 64(7):3-9 (1967).

Sharrard, W. J. W., "The distribution of the permanent paralysis in the lower limb in poliomyelitis," J. Bone Joint Surg. 37B: 540-558 (1955).

Sherrington, C. (1906), The Integrative Activity of the Nervous System [in Russian], Leningrad, Science Publishing House, 1969.

Sherrington, C. S., and Laslett, E. E., "Observations on some spinal reflexes and the interconnection of spinal segments," J. Physiol. (England) 29:58-96 (1903).

Slaughter, D. G., Nashold, B. S., and Somjen, G. G., "Electrical recording with micro- and macroelectrodes from the cerebellum of man," J. Neurosurg. 33:524-528 (1970).

Somjen, G., Carpenter, D. O., and Henneman, E., "Responses of motoneurons of different sizes to graded stimulation of supraspinal centers of the brain," J. Neurophysiol. 28:958-965 (1965).

Sotgiu, M. L., and Cesa-Bianchi, M. G., "Primary afferent depolarization in the cuneate nucleus induced by stimulation of cerebellar and thalamic nonspecific nuclei," EEG Clin. Neurophysiol. 29:156-165 (1970).

Sperry, R. W., "Neural basis of the spontaneous optokinetic response produced by visual inversion," J. Comp. Physiol. Psychol. 43: 482-489 (1950).

Spiegel, E. A., Der Tonus der Skelettmuskulatur, Berlin, 1927.

Spielberg, P. I., "Electric brain and muscle potentials in man during voluntary movements," Fiziol. Zh. SSSR im. I. M. Sechenova 30:546-551 (1941).

Sprague, J. M., and Chambers, W. W., "Regulation of posture in intact and decerebrate cat. I. Cerebellum, reticular formation, vestibular nuclei," J. Neurophysiol. 16:451-463 (1953).

Sprague, J. M., and Chambers, W. W., "Control of posture by reticular formation and cerebellum in the intact, anesthetized and unanesthetized and in the decerebrated cat," Am. J. Physiol. 176:52-64 (1954).

Sprague, J. M., Schreiner, L. H., Lindsley, D. B., and Magoun, H. W., "Reticulo-spinal influences on stretch reflexes," J. Neurophysiol. 11:501-507 (1948).

Stewart, D. H., and Preston, J. B., "Functional coupling between the pyramidal tract and segmental motoneurons in cat and primate," J. Neurophysiol. 30:453-465 (1967).

Struppler, A., and Struppler, E., "Neurophysiologische Untersuchungen an de-afferentierten Muskeln des Menschen," Z. Biol. 111:438-448 (1960).

Sverdlov, S. M., and Maksimova, E. V., "Afferent control in the presence of the action of pyramidal impulses on spinal motoneurons," Fiziol. Zh. SSSR im. I. M. Sechenova 52:441-446 (1966).

Sverdlov, S. M., and Maksimova, E. V., "The pyramidal and extrapyramidal motor systems," in: The General and Specific Physiology of the Nervous System [in Russian], Leningrad, Science Publishing House, 1969, pp. 338-361.

Sverdlov, Yu. S., "Long-lasting changes in the amplitude of monosynaptic extensor reflexes evoked by afferent volleys in somatic nerves," Fiziol. Zh. SSSR im. I. M. Sechenova 53:1414-1423 (1967).

Sweet, J. E., and Bourassa, C. M., "Short latency activation of pyramidal tract cells by group I afferent volleys in the cat," J. Physiol. (England) 189:101-117 (1967).

Szenthagothai, J., "Die zentrale Innervation der Augenbewegungen," Arch. Psychiatr. Nervenkr. 116:721-760 (1943).

Táboříkova, H., "Fraction of the motoneurone pool activated in the monosynaptic H-reflexes in man," Nature (London) 209:206-207 (1966).

Tanaka, R., "Activation of reciprocal Ia inhibitory pathway during voluntary motor performance in man," Brain Res. 43:649-652 (1972).

Teasdall, R. D., Park, A. M., Languth, H. W., and Magladery, J. W., "Electrophysiological studies of nerve and reflex activity in patients with lesions of the nervous system. II. Disclosure of normally suppressed monosynaptic reflex discharge of spinal motoneurones by lesions of lower brainstem and spinal cord," Bull. Johns Hopkins Hosp. 91:245-256 (1952).

Teuber, H. L., "Alterations of perception after brain injury," in: Brain and Conscious Experience (J. C. Eccles, ed.), New York, Springer-Verlag, 1966, pp. 182-216.

Thach, W. T., "Discharge of Purkinje and cerebellar nuclear neurons during rapidly alternating arm movements in the monkey," J. Neurophysiol. 31:785-797 (1968).

Thach, W. T., "Discharge of cerebellar neurons related to two maintained postures and two prompt movements. I. Nuclear cell output," J. Neurophysiol. 33:527-536 (1970a).

Thach, W. T., "Discharge of cerebellar neurons related to two maintained postures and two prompt movements. II. Purkinje cell output and input," J. Neurophysiol. 33:537-547 (1970b).

Thach, W. T., "Cerebellar output: Properties, synthesis and uses," Brain Res. 40:89-97 (1972).

Thomas, R. C., and Wilson, V. J., "Recurrent interactions between motoneurons of known location in the cervical cord of the cat," J. Neurophysiol. 30:661-674 (1967).

Totsuka, G., Suzuki, M., and Kubota, K., "Vestibular influence on the gamma-efferent motor system observed by angular acceleration," Acta Oto-Laryngol. 179:18-24 (1963).

Tower, S. S., "Pyramidal lesion in the monkey," Brain 63:36-90 (1940).

Twitchell, T. E., "Sensory factors in purposive movement," J. Neurophysiol. 17:239-252 (1954).

Tzekov, Tr., "Correlation between the reaction time and the amplitude of the H-reflex of the muscle involved in the response," Agressologie 13D:25-30 (1972).

Uemura, K., and Preston, J. B., "Comparison of motor cortex influences upon various hind-limb motoneurones in pyramidal cats and primates," J. Neurophysiol. 28:398-412 (1965).

Ukhtomskii, A. A., "On the dependence of cortical motor effects on secondary central influences," Proceedings of Imp. St. Petersburg Society of Naturalists, 41(Issue 2):177-392 (1911).

Ukhtomskii, A. A., "The Dominance Principle," in: Collected Essays [in Russian], Vol. 1, Leningrad State University Press, 1950, pp. 197-201.

Ukhtomskii, A. A., "An outline of the physiology of the nervous system," in: Collected Essays [in Russian], Vol. IV, Leningrad State University Press, 1954.

Uznadze, D. N., Experimental Basis of the Psychology of Adjustment [in Russian], Tbilisi, 1961.

Valbo, A. B., "Muscle spindle response at the onset of isometric voluntary contractions in man. Time difference between fusimotor and skeletomotor effects," J. Physiol. (England) 218:405-431 (1971).

Vasilenko, D. A., and Kostyuk, P. G., "Characteristics of the activation of different spinal neuronal groups on stimulation of the cat sensorimotor cortex," Zh. Vyssh. Nervn. Deyat. im. I. P. Pavlova 15:695-704 (1965).

Vasilenko, D. A., and Kostyuk, P. G., "The phasic and tonic components of pyramidal influences on motoneurons," Dokl. Akad. Nauk SSSR 169:731-734 (1966).

Vasilenko, D. A., and Vucho, I., "Synaptic processes in lumbar motoneurons on stimulation of the sensorimotor cortex," Zh. Vyssh. Nervn. Deyat. im. I. P. Pavlova 16:52-61 (1966).

Vasilevskii, N. N., Neuronal Mechanisms of the Cerebral Cortex [in Russian], Leningrad, Medicine Publishing House, 1968.

Vaughan, H. G., Jr., and Costa, L. D., "Analysis of electroencephalographic correlates of human sensorimotor processes," EEG Clin. Neurophysiol. 24:288-289 (1968).

Vaughan, H. G., Jr., Costa, L. D., and Ritter, W., "Topography of the human motor potential," EEG Clin. Neurophysiol. 25:1-10 (1968).

Vaughan, H. G., Jr., Gross, E. G., and Bossom, J., "Cortical motor potential in monkeys before and after upper limb deafferentation," Exp. Neurol. 26:253-270 (1970).

Veber, N. V., "Electrical reactions of the anterior cord roots evoked by stimulation of the pyramidal tract after deafferentation," in: Mechanisms of Compensative Adaptations [in Russian], Moscow, Science Publishing House, pp. 87-98, 1964.

Vedel, P., "Mise en evidence d'un controle corticale de l'activite des fibres fusimotorices dynamiques chez le chat par la voie pyramidal," C. R. Acad. Sci. Paris 262:908-911 (1966).

Vinogradov, M. I., and Shvang, L. I., "On the formation of a warning reaction during sustained muscular contraction," in: Problems of the Physiology of the Central Nervous System [in Russian], Moscow-Leningrad, pp. 137-140, 1957.

Voronin, L. L., and Tanenholz [Tanengol'ts], L. I., "The interaction of neurons of different modalities at neurons of the motor area of the cortex," in: The Organization of Interneuronal Connections [in Russian], Moscow, pp. 11-26, 1967.

Vučo, J., and Todorovič, B., "Postishemicna potencijacija alfa tonusnih motornih neurona," Arhiv Biol. Nauka 16(1-2):1-13 (1964).

Vvedenskii, N. E. (1897), "On the interrelations between psychomotor centers," in: The Physiology of the Nervous System [in Russian], 3rd Ed., Vol. 1, Moscow, State Medical Publishing House, 1952, pp. 181-188.

Vvedenskii, N. E. (1913), Lecture course on human and animal physiology conducted at the University of Petersburg from 1911 to 1913; Complete collection of papers, 5 volumes pub. Leningrad, 1954.

Vvedenskii, N. E., and Ukhtomskii, A. A. (1909), "Reflexes of antagonistic muscles on electrical stimulation of a sensitive nerve," in: The Physiology of the Nervous System [in Russian], 3rd Ed., Vol. 1, Moscow, State Medical Publishing House, pp. 234-259, 1952.

Wachholder, K., and Altenburger, H., "Über die Wechselbeziehungen zwischen den Sehnenreflexen und der antagonistischen Innervation unserer Muskeln," Pfluegers Arch. Gesamte Physiol. Menschen Tiere 203:620-631 (1924).

Wachholder, K., and Altenburger, H., "Beiträge zur Physiologie der willkürlichen Bewegung. XI. Über die Genese der Antagonistentätigkeit," Pfluegers Arch. Gesamte Physiol. Menschen Tiere 215:622-626 (1927).

Wagner, R., "Über die Zusammenarbeit der Antagonisten bei der Willkürbewegung. II. Abhängigkeit von mechanischem Bedingungen," Z. Biol. 83:59-93 (1925).

Wall, P. D., "Presynaptic control of impulses at the first central synapse in the cutaneous pathway," Prog. Brain Res. 12:92-118 (1964).

Wall, P. D., "The laminar organization of dorsal horn and effects of descending impulses," J. Physiol. (England) 188:403-423 (1967).

Walter, W. G., "Slow potential waves in the human brain associated with expectancy, attention and decision," Arch. Psychiatr. Nervenkr. 206:309-316 (1964).

Wiesendanger, M., "The pyramidal tract. Recent investigations on its morphology and function," Ergeb. Physiol. 61:72-136 (1969).

Wiesendanger, M., and Laursen, A. M., "Pyramidal influences on gamma- and alpha-motoneurones in the cat," in: Muscular Afferents and Motor Control, Nobel Symposium (R. Granit et al., eds.), Stockholm, 1966, pp. 465-466.

Willis, W. D., Tate, G. W., Ashworth, R. D., and Willis, J. C., "Monosynaptic excitation of motoneurons of individual forelimb muscles," J. Neurophysiol. 29:410-424 (1966).

Wilson, S. A. K., Modern Problems in Neurology, New York, William Wood and Co., 1929.

Wilson, V. J., "Post-tetanic potentiation of polysynaptic reflexes of the spinal cord," J. Gen. Physiol. 39:197-206 (1955).

Wilson, V. J., "Regulation and function of Renshaw cell discharge," in: Muscular Afferents and Motor Control, Nobel Symposium (R. Granit et al., eds.), Stockholm, 1966, pp. 317-329.

Wilson, V. J., and Kato, M., "Excitation of extensor motoneurons by group II afferent fibers in ipsilateral muscle nerves," J. Neurophysiol. 28:545-554 (1965).

Wilson, V. J., and Yoshida, M., "Comparison of effects of stimulation of Deiters nucleus and medial longitudinal fasciculus on neck, forelimb and hindlimb motoneurons," J. Neurophysiol. 32: 743-758 (1969).

Wilson, V. J., Talbot, W. H., and Kato, M., "Inhibitory convergence upon Renshaw cells," J. Neurophysiol. 27:1063-1080 (1964).

Woolsey, C. N., "Organization of somatic sensory and motor areas of the cerebral cortex," in: Biological and Biochemical Bases of Behavior, Madison, University of Wisconsin Press, 1958, pp. 63-82.

Woolsey, C. N., Travis, A. M., Barnard, J. W., and Ostenso, R. S., "Motor representation in the postcentral gyrus after a chronic ablation of precentral and supplementary motor areas," Fed. Proc. Fed. Am. Soc. Exp. Biol. 12:160 (1953).

Yankovska, E., and Gurska, T., "The development of a scratch reflex in the deafferented cat and rat limb," in: Central and Peripheral Mechanisms of Motor Activity in Animals [in Russian], Moscow, Science Publishing House, 1960, pp. 248-249.

Yokota, T., and Voorhoeve, P. E., "Pyramidal control of fusimotor neurons supplying extensor muscles in the cat's forelimb," Exp. Brain Res. 9:96-115 (1969).

Yoshida, M., Yajima, K., and Uno, M., "Different activation of the two types of the pyramidal tract neurons through the cerebellothalamocortical pathway," Experientia 22:331-332 (1966).

Zadorozhnyi, D. A., Vasilenko, D. A., and Kostyuk, P. G., "Pyramidal influences in the interneurons of cat segmental reflex arcs," Neirofiziologiya 2:17-25 (1970).

Zalkind, M. S., "On the non-specific components of local motor skills in man," in: 2nd International Seminar on the Regulations of Movement, Sofiya, 1972, p. 21.

Zalkind, M. S., and Shlykov, V. Yu., "On generalized motor reactions in man," Neirofiziologiya 6:19-25 (1974).

Zhukov, V. I., "Electrophysiological analysis of the mechanisms of the reciprocal interaction of antagonist muscles in man" (Master's thesis), Moscow, 1971.

INDEX